Breeding Shropshire Sheep

by F.M. Chandler

with an introduction by Jackson Chambers

Self Reliance Books

Get more historic titles on animal and stock breeding, gardening and old fashioned skills by visiting us at:

Introduction

I am pleased to present yet another practical title on breeding and raising livestock.

The work is in the Public Domain and is re-printed here in accordance with Federal Laws.

As with all reprinted books of this age that are intended to perfectly reproduce the original edition, considerable pains and effort had to be undertaken to correct fading and sometimes outright damage to existing proofs of this title. At times, this task is quite monumental, requiring an almost total "rebuilding" of some pages from digital proofs of multiple copies. Despite this, imperfections still sometimes exist in the final proof and may detract from the visual appearance of the text.

I hope you enjoy reading this book as much as I enjoyed making it available to readers again.

Jackson Chambers

CHAMPION SHROPSHIRE FLOCK, INTERNATIONAL SHOW, CHICAGO, 1909.

Also winners of the Silver Cup for best five Shropshires either sex any age, and with the First Prize ram lamb they won the Silver Cup for Champion Flock consisting of one ram one year old or over, one ram lamb, two ewes one year old or over and two ewe lambs, thus being the undefeated Shropshire group of America. Speaking of the Shropshire exhibit at the great 1909 International that high authority *Breeders Gazette* says "this popular breed of 'farmers sheep' sent forward its strongest showing ever made in America. It was a hard fought contest from start to finish." It was the first time ever that all the present big breeders from England, have come together fully prepared for the strongest show on the Continent. The above Champion Flock was owned and exhibited by Chandler Bros., "Clover Hill Farm," Chariton, Iowa.

The Farm and the Flock

Farmers are working under different conditions than did their forefathers—in many instances the soil has been tilled until it is weakened. The agricultural population realizes more fully than ever before the necessity of restoring strength to worn-out farms. also of keeping up even those that are the most productive. Farmers do not desire to repeat the past folly of continually raising grain and not returning anything to the soil but are anxious to learn the best solution of this problem. Land is getting higher and higher in price and not only are owners of deteriorated farms striving to bring theirs up to an average but those who have the most valuable and richest farms want some means of maintaining this high standard and deriving the necessary profit from a large investment. Therefore the main purpose of the farmer is to increase the productive power of the soil and to raise upon that land what will make the largest net returns. As no soil can be continually farmed for grain it must be changed to clovers and other grasses and then comes the question, "What is it that will give the best results in increasing soil fertility and also the largest profits from grasses, both green and in the form of hay?" The fact that many are learning the correct answer to this question is one reason why the sheep business increasing generally. Their droppings are the richest of known natural fertilizers and are well scattered over the pastures. In addition to this, the flock is the greatest of weed destroyers and killing such large quantities of numerous weeds preserves in the soil that plant food which the weeds would have consumed. So the flock adds strength to the soil in two ways, and the good results from their eating nearly every known weed is invariably underestimated. No other domestic animal will so completely clear the farm, and at the same time sheep use weeds as food. Apart from preserving soil fertility. the total riddance of weeds adds much to the appearance of land. Farmers, as a whole, have partially learned the value of a flock in this respect and those who are working to preserve their farms are not scoffed at as they were a few years ago. As more attention is given to the soil, the number of flocks will increase. If sheep consumed as much grain and hay accordingly as other stock, sold for the same market price, and had no wool, flocks would anyway eventually become more numerous on our farms in order to obtain the results just mentioned. But our population must be clothed and their average

3

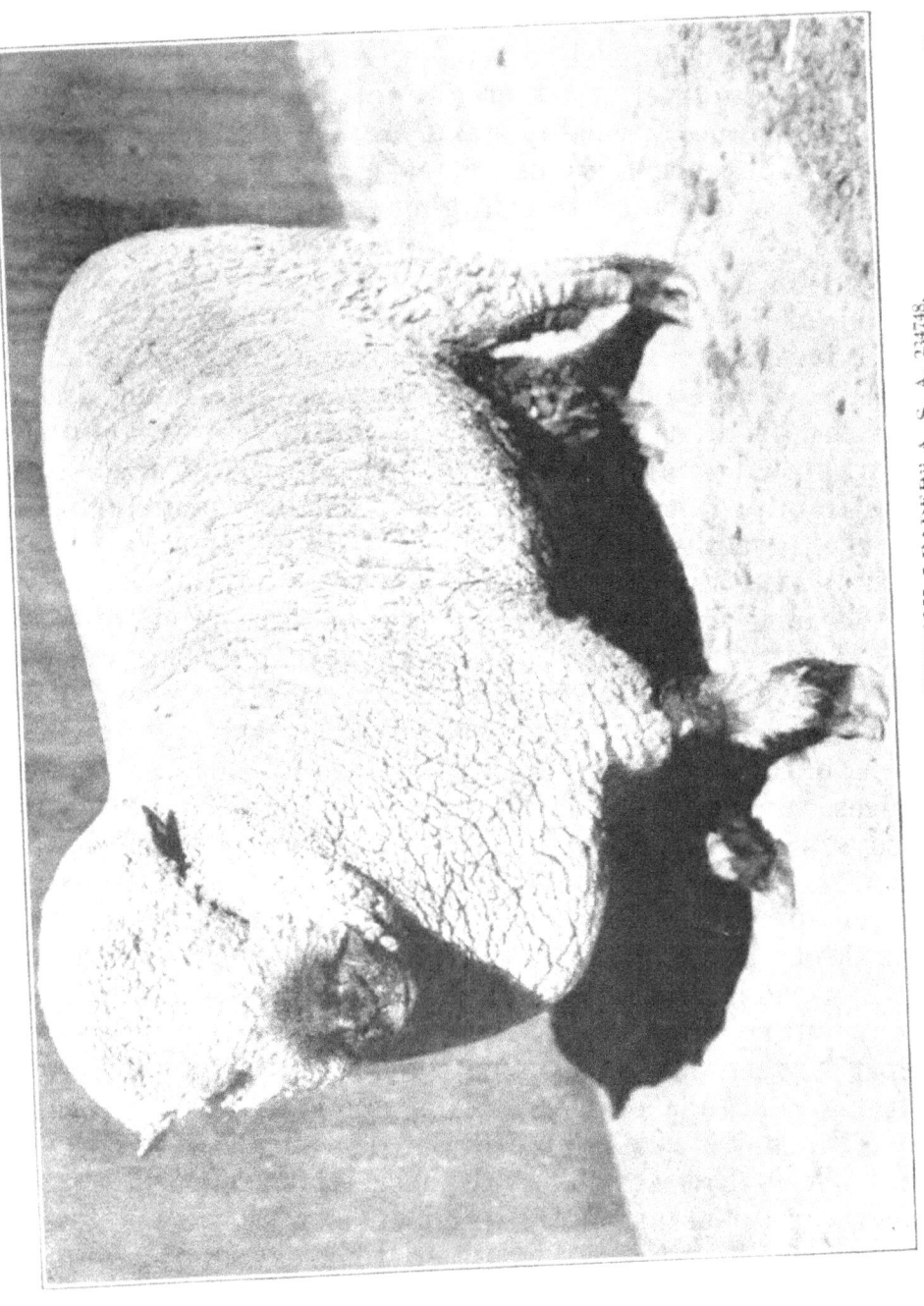

"CLOVER HILL WHITTINGTON LEADER" A. S. A. 23748. Exhibited by Chandler Bros., First Prize two-year old and Champion Shropshire ram any age at State and National Fairs 1906. Proved to be a very impressive sire and was used three years in the "Clover Hill" Flock. "Clover Hill Farm," Chariton, Iowa.

wealth is such that good clothes will be purchased—that means large demand for wool and it is in evidence according to the high price of wool. The strong continual demand will always keep it there, too, just the same as wheat, corn, etc., which the millions of people must have. The wool clip will invariably fully pay for the yearly upkeep of the flock, and no other domestc animal has a "side product" that will pay its yearly board bill. The lamb crop comes in as clear profit and is a large return in comparison with the investment. More mutton is being consumed per capita and the great increase in population has made a noticeable advancement in the demand for mutton. The price of lambs on the leading markets during recent years has averaged higher than cattle or hogs. Even if prices had been equal, lambs would have been the most profitable owing to low cost of production. It has been demonstrated that from a given amount of feed lambs will make the largest gain, and they are also much easier cared for than other stock. Many farmers have been born where cattle, hogs, and corn were about all they saw, and truly good returns have come from that sort of farming but it cannot always be continued. Experienced men say that the profits are not nearly so great now as in the past and if it were continued without variation the farms would not be as valuable as they might have been. The necessity of a change is realized and nothing else fills the place like a flock of sheep. Evidence of this comes from the large number of flocks which have been founded during recent years on just such farms. Years ago the prevalent idea was that sheep were only good for rough, brushy land which could not be plowed. They did give the largest obtainable returns from such land, but now farmers also know that sheep in their place give the largest returns from high priced land. Those who realize that no land can raise corn for an indefinite period are in a majority of cases putting in a flock of sheep. The principal cause of less flocks seems to have been because most farmers did not grow up where sheep were kept, so that they have never given any attention to the true value of a flock. As deeper study is given to sheep, the fewer will be the number of farms without them. English farmers have long ago learned that in order to derive the greatest possible profit from a farm, a flock of sheep must be kept upon it. As American land approaches the value of theirs, the absolute necessity of soil fertility comes into prominence, and farmers figure for the last dollar that their farms will produce, either directly or indirectly, then sheep will come into their proper place and there will be the right relation between the farm and the flock.

The Cost of Producing Mutton

As farming cannot be successfully continued without occasional change to grasses for the maintenance of live-stock thus fertilizing the soil in different ways and owing to the fact that sheep make the best use of all odds and ends about the pastures, meadows, rye patches, corn with rape underneath, etc., and are the best of live stock to fertilize by their droppings, no farm however rich in natural fertility or high in price will give largest net returns without a flock of sheep. Therefore the question is "What breed of sheep gives the largest profits continually?" Shropshires are the most economical producers of mutton, giving higher returns in carcass weight for food consumed than any other of the acknowledged mutton breeds. Each pound of mutton they produce is worth more money than the coarse-grained sort from the extremely large mutton breeds. Each year the range in market price is getting wider and wider between the compact, firmly and evenly-fleshed Shropshire lambs which give quality carcasses of handy weights, and the larger rough breeds which give less dressed percentage of mutton which is also of much lower quality. From a current issue of probably the most reliable publication regarding Chicago live-stock markets we quote the following:

"It must not be presumed that all lambs are realizing lofty prices. Only high dressers are equal to the performance and dressed meat percentages are closely watched. A band of shorn lambs costing $8.00 on the hoof actually made dearer mutton on the hooks by $1.00 per hundred weight than another purchase costing $8.50 alive."

Buyers for the large killers and packers are now-a-days close observers of how every purchase dresses out in quality and weight on the hooks. In future years even a closer discrimination will be made against lambs which do not "kill well." Returns are being kept close tab on in order that lambs and sheep will be purchased according to their real value—no guess-work about it. Notice in the market reports from time to time that medium-weight quality lambs bring nearly double the price that coarse fellows with poor quality do. When you go to the butcher's shop do you want a chunk of coarse-grained, fatty mutton? If so you are one in a thousand because the other 999 will want a rich lean piece fine as possible in quality. People now-a-days know the difference in taste between the two, and place all preference for that which is fine in grain. Lowest actual cost of production per pound of mutton and the very highest price when sold is certainly making the former indifferent sheep raisers "turn the tables" and keep the breed which is really the most profitable when everything is taken into consideration. Sheep raisers are also particularly noticing how much lower the annual cost of maintaining a flock of Shropshire breeding ewes is than those of any other breed. Not only is it important to have the class of lambs which make good gains and command highest price, but it is most desirable to materially lower the cost of first producing those lambs. Ewes of other breeds require a larger amount of green food and some grain in addition while Shropshire ewes will be nursing fat lambs and in perfect condition on a rougher, poorer pasture and without grain. Many times at the same season of the year have the writers visited breeding flocks of the Shropshire and various other breeds in different sections of both America and England and noticed the conditions exactly as stated above. Shropshires are easiest to keep in thrifty condition and ofttimes at practically half the cost of the upkeep of flocks of other breeds. Shropshires are naturally good feeders and exceptionally strong in constitution, having the inherent robustness of their origin from the hill breeds of the English county from whence the Shropshire breed takes its name. Strength of constitution is a prime requisite in all breeding or feeding sheep. In the life of animals things come up as various sorts of trouble and hardship which must be withstood by the animal system, the weak constituted ones suffering to a greater or less extent under these conditions but those with strong constitutions ward off the trouble and are hale and hearty. The strong-constituted sheep possesses the highest degree of digestive and assimilative power and even under unfavorable conditions makes most thorough use of all its food. Shropshire fleeces have the greatest density and length combined thus giving heavy weights and complete protection, altogether making what might well be termed the unequalled general-purpose sheep for the farmer and breeder. The Shropshire fleece is a perfect covering all over and under the body and is bred that way not only for increased weight of wool but for absolute protection from damp, cold weather and storms. The general farmer needs such a breed which will not be soaked to the skin when there is rain or blowing snow. The dense Shropshire fleece together with their strong constitution insures health and vigor under all conditions and in varying climates. These characteristics by keeping up the highest degree of thrift aid the sheep in making largest gains thereby lowering the cost of Shropshire mutton production. On the average farm wherever located the Shopshire will give best possible results in the economical production of highest-class mutton.

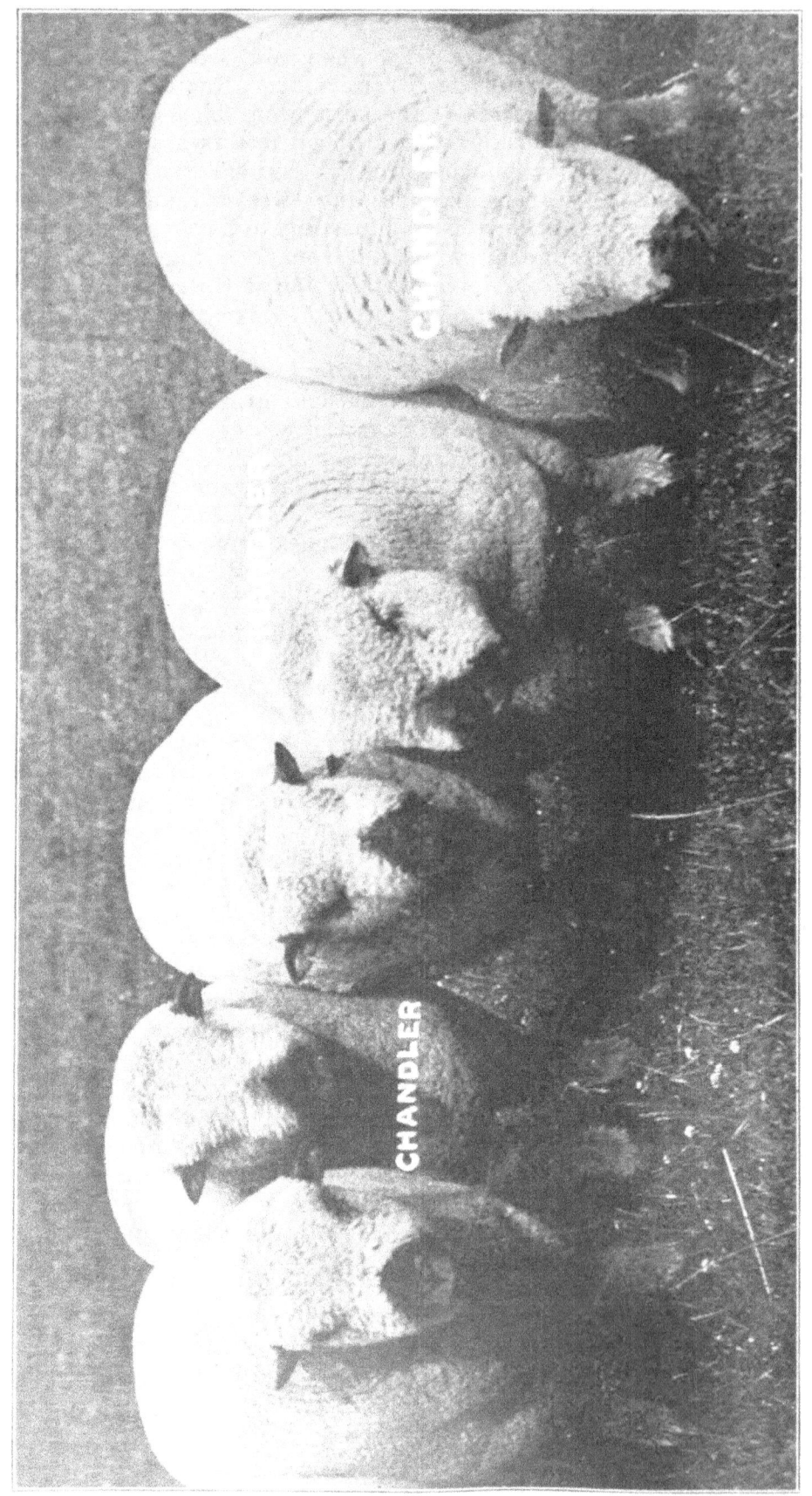

"Clover Hill's 4356"   "Clover Hill's 4357"   "Clover Hill's 4377"   "Clover Hill's 4359"   "Clover Hill's 4275"
        (297513)   (297514)   (297510)   (297516)   (297511)
First Prize Pen Five Shropshire Rams any age (Iowa-bred) Iowa State Fair 1909. These are all yearlings, bred and exhibited by
                        Chandler Bros., "Clover Hill Farm," Chariton, Iowa.

Selecting Flock Headers for Grade Flocks

Careful methods of breeding have placed the Shropshire breed of sheep in its present possession of desirable qualities. The greatest profits come to the best breeders. The common class of lambs does not attract the same number of purchasers as the better lot, even though the price asked for the former is much lower. This superiority can be attained only by proper methods of breeding. Although the ewe portion of the flock plays a very important part in breeding, the greatest, cheapest and quickest results can be obtained by using the highest class of registered rams. The ram makes a mark on every lamb while the ewe affects but one or two, as the case may be.

Before commencing your look for a flock-header, get it thoroughly into your mind just the sort that will do your flock the most good, and after that do not stop until you get him. If you have a grade flock and are producing lambs for market purposes, remember that the strong-constitutioned thick-fleshed fellows top the market, and your bunch of lambs must be uniform in size, type, and density of wool to present the desired appearance in the sale pen. Constitution is a prime requisite in sheep, whether they are for the breeding pen, feed lot, or showyard. The extended nostril, strong short neck, wide deep chest, and well-sprung rib all indicate that the heart, lungs and digestive organs have plenty of room in which to do their proper and required work.

Get a strong-constitutioned ram, correct in mutton conformation, with a dense clear fleece, and all the size possible. The bigger the sheep the better so long as he has good quality of flesh. There is a "happy medium" which combines much size with quality—that is the right sort. It doesn't pay to raise coarse poor feeders or those that fatten in patches, because the market doesn't want that sort, but it never pays to raise little bits of things either. Breed for all the size you can possibly combine with quality. The butcher wants the lamb which will dress out the largest percentage of natural flesh in the most valuable cuts, the back and hind-quarters, and you need not think that a bunch of bare-backed, narrow hind-ended lambs would go through any leading market at the same price that the straight-backed wide thickly fleshed sort would.

When purchasing rams remember that in addition to being the hardiest finest qualitied mutton breed for the farmer Shropshires are the heaviest shearers of the Down mutton breeds.

"Clover Hill's 6004"   "Clover Hill's 8"   "Clover Hill's 34"   "Clover Hill's 27"   "Clover Hill's 12"
(289092)   (289076)   (298882)   (298889)   (298892)
First Prize Pen Five Shropshire Rams any age (open to the world competition) Iowa State Fair 1909.   All yearlings, owned and exhibited by
Chandler Bros., "Clover Hill Farm," Chariton, Iowa.

The dense Shropshire fleece adds to the general thrift of the flock by affording natural complete protection from bad weather and also greatly increases the annual income. From old Shropshire flocks which have been bred for heavy fleeces, rams can be obtained which shear 15 lbs. and upwards of clear white dense wool. To make the largest returns from the flock investment no ram should be used which shears less than 15 lbs. per clip. A large income can be obtained from the wool by careful breeding and not sacrifice anything in mutton conformation.

Decided masculinity is required in the ram, this being indicated by general burliness of the head, thickness of the neck, and general massiveness with a bold assertive appearance. A ram with a narrow nose and head, a long slim neck, and lacking in vigor, rarely if ever, was known to be an impressive sire. If you are accustomed to purchasing a ram of the medium class, the additional $5 required for the purchase of a superior one may seem to be quite a large sum, but a small investment in this way nearly always results in a much larger future income from the flock. Those who have used good rams never turn back to using the more common class, and if you get a better ram this year than ever before, your flock will commence to make larger and better returns and make breeding more interesting.

Best value in sires is obtained by purchasing from the oldest reputable flock. By introducing such individuality and breeding which has taken the ram breeder a life-time to produce, you are at a comparatively small cost raising your flock to a high standard. Carefully bred rams from a good old flock are, by reason of the superiority of ancestors, reliable sires which will most strongly impress their good qualities upon the offspring. One of the most profitable results coming from the use of such rams on grade flocks is that the lamb crop is so uniform in type, markings, conformation of body, and density of fleece. Big rams from a flock which has long been bred for size will sire a most uniform lot of large individual lambs. There is great difference between the breeding results from a ram obtained from a really superior flock and another obtained from a more common registered flock. The broad-minded successful breeder is also a better man to do business with and experience has taught him how to meet customers on a fair basis. It is great satisfaction to have relation with breeders whom you know can be depended upon. Their sheep are the highest class, pedigrees correct, prices in accordance with actual value, and all dealings are handled in a business-like manner.

(Read in this connection the fifth article in the book after this one.)

Advancement in American Pure-Bred Shropshire Flocks

The improvement made in American pure-bred flocks during recent years has been great, and the industry possesses very marked stability. There are many causes leading up to this, among the most important being the spreading out in production of mutton sheep, thereby causing competition and creating a demand for high-class pure-bred sires. Some years ago there was not such close discrimination in market centers against the common lamb as there is today, nor was such a premium paid for the really good ones. General farmers and ranchmen have learned that sheep are a valuable asset to their possessions, and are now making a study and business of mutton production, where it formerly was thought of lightly. There is a consumer's demand for mutton far greater than that of the past, and these facts make the sheep industry one which is recognized by all who are interested in the production of live stock.

Competition on the markets has assisted in making the wide variance in price between "choice" and "common" lambs. Those men who have marketed a few bunches of lambs have had this difference in price deeply stamped upon their minds, and are now striving to produce the sort that will top the market. To bring on this better class of mutton the producer has not only had to feed well, but also improve his methods of breeding. He has purchased a good class of registered rams that have sired a more uniform lot of thick-fleshed lambs than were formerly reared. Mutton producers have learned of the profits to be derived therefrom, and profits are what are worked for. The demand for such rams has taught pure-bred breeders that it is now the best or none, and the ordinary rams will not sell at any price. Some of America's best pure-bred breeders are also to be congratulated on their efforts to help bring about this condition, because by their sending out none other than strictly first-class sires they have been of great assistance in convincing the mutton producer that the best rams give the greatest net profits.

The market's paying a premium for top lambs, the few veteran breeders sending out none but high-class sires, and the good results which the mutton-producer has really experienced from the use of such are all factors in the expansion of breeding pure-bred sheep. The past has given experience to the mutton-producer that the use of good sires is profitable, and to the ram breeder that he cannot sell the common class of males. Discrimination on the market against cull lambs makes the producer as a ram buyer discriminate against culls— and it is a good thing. The day of the grade ram is of the past, and the use of low-class registered rams is going rapidly in the same direction—our wish being that it may gain increased rapidity.

The careless breeder of pure-breds must change his ways or suffer loss, the same as the producer of the common sort of mutton lambs which are a drug on any market. These facts have been so thoroughly demonstrated that a great change has taken and is taking place. The aim of a vast majority of pure-bred breeders is to produce the very best that is possible. Especially within the last 20 years have importations of select breeding material been made from England, and it has given the breeders a solid foundation to work on and enabled us to make much more rapid advancement than would have otherwise been possible. Breeders are producing such excellent specimens of the Shropshire breed that it is none but Great Britain's very tops that will excel ours. This should be a stimulant and a matter of great satisfaction to our breeders, and we believe it is.

The competition among pure-bred flocks is keen, and a majority are striving to see who can make the greatest possible improvement. This enables most mutton-producers to obtain the desired quality in sires, thereby raising the general standard of the sheep industry. Yet there are a number of grade and range flocks upon which a common class of rams is used, but a study of the matter reveals the fact that in most cases it is because the owners are unable to obtain the class of rams they really want. The pure-bred business is at a high state of perfection in many sections of America, but it needs general expansion by having more energetic breeders in new territory. More breeders who have grade flocks should purchase a select little foundation of pure-breds, and by gradually working into the business in that way they will derive great personal benefit and help increase the quality of the nation's sheep industry.

A visitor in the sheep pens at any of the state fairs will be favorably impressed with the excellence which pure-bred Shropshires possess, and this will be even increased in the future. The market requirements will be for better lambs, the producer will purchase none but first-class rams, and the pure-bred breeder must have the best in the land. These facts demonstrate the great advancement in the sheep industry, and point to the requirement of pure-bred breeders that they exercise the best possible judgment in improving the nation's flocks.

"CLOVER HILL KING" A. S. A. 256751
A sire now in service in the "Clover Hill" flock of Shropshires.

Favorable Points of the Pure-Bred Shropshire

Size—When it comes to getting the greatest size and highest quality combined, the Shropshire stands in a class of its own. At "Clover Hill Farm" we have had rams weigh up to 325 lbs., shear 22 lbs., and possess all the quality of the small mutton breeds.

Constitution—Shropshires are naturally strong in constitution, having the inherent robustness of their origin from the Hill breeds. Shropshires have thrived exceptionally well in nearly all parts of the world through extremes of wet, dry, heat and cold, on high and low land.

Ewes Good Mothers—Under ordinary conditions Shropshires will raise a larger number of desirable lambs than any other breed. In one instance lately we saw a Shropshire ewe with 4 big, strong lambs which she had given birth to 6 weeks previous. Shropshire ewes are excellent nurses and lambs are born strong. Usually at least half of the ewes drop twins.

Wool—Shropshires are covered all over and under the body with a heavy, dense fleece of good length, insuring absolute protection from storms and weighing a usual average of from 9 to 15 lbs., which brings very top market price, owing to quality and strength.

Mutton—Shropshires are compact, even and firm fleshed and give the highest dressed percentage of quality carcasses of lean, juicy meat, thus bringing extreme top price. The quality of Shropshire mutton has been a main factor in raising American mutton trade to its high standard by offering a better product to the consumer.

Cost of Production and Early Maturity—Shropshire breeding flocks are easiest to keep in thrifty condition and the breed stands out pre-eminent as economical producers of mutton par excellence. Shropshires being naturally strong-constitutioned possess the highest degree of digestive and assimilative power, thus making most thorough use of all their food. With these inherent qualities, together with extreme quality and classy character, Shropshire lambs are ready for market at an early age.

Impressive Sires—A superior feature of the pure Shropshire ram is that he transmits these good qualities in a marked degree to his offspring. Shropshire rams can be depended upon to get the most profitable class of lambs when used on grade or ewes of other breeds. The large size, strength of constitution, conformation, quality of mutton, black face, and attractive general character being exclusive Shropshire characteristics are always transmitted to cross-breds. Altogether the Shropshire is an unequalled general-purpose sheep for the farmer and breeder.

One Breed and One Only

In order to achieve full success in any business a man must specialize to concentrate his thoughts and efforts upon one thing. If there are two or three subjects which demand attention, he is weakened in all of them and as the world to-day demands first-class services a person must do one thing and do it well. By not spreading out, your whole time will then be devoted to one thing and your knowledge of that subject becomes superior and you can give customers the best possible service. Even more than in a commercial business do these facts apply to the sheep breeder. The commercial man purchases his wares of a certain quality and although placed on the shelves they will keep up to that standard. All he has to do is to sell them. The sheep breeder after getting his foundation flock, has a continual study of how he can make improvement by proper mating and by good feeding also assist Nature in the full development of the lambs. Sheep which the pure-bred breeder produces are not eaten for food but are offered for the sole purpose of improving other flocks. The breeder in looking to the future readily realizes that not only must the sheep he sells appear well at the time of sale but they must possess such breeding and have been reared so they will give the purchaser uniformly desirable results. Therefore the study and management being required to bring one breed of sheep to success for you is much greater than is required from the commercial man to succeed with one line of goods. But the most successful of them are those that have "specialized" and the most successful sheepmen of the world are those who stick to one breed, and one only. The quality of sheep sent out and general services given are what gain reputation. The good reputation is a result of having given to your customers the kind of treatment that profited them. After you have selected the one breed of sheep that you like, study and work to make the greatest success attainable so in time the sheepmen all over the continent will have learned that for that one particular breed you can give them better services than can be obtained elsewhere and that sheep from your flock give the best possible results as breeders. You are then gaining success in its strictest meaning. Is it real success or can satisfaction be derived from breeding two, three, or more different breeds and just making a sale wherever you get a chance?

Then the old saying will come up, "Jack of all trades and Master of none." Could you, in that position, give customers for different breeds as good service or as reliable sheep for the breeding flock as you could have given with but one breed? The only answer is "NO." A review of the past and present sheep business verifies such a universal answer, too, because, as a rule, those who have attempted to branch out and produce more than one breed have not given service which is up to the ideal of the best sheepmen. The pure-bred breeder with one breed is filled with spirit and desires to only send out such specimens as will be a credit to the breed and his flock. Make your motto, "One breed and one only—Success with that." In the end the financial returns will be greater, you will have the full satisfaction of being a first-class breeder of a superior flock, and will be considered a solid rock in the foundation for an improved sheep industry. The country wants more breeders who produce superior sheep that when sold will continue to make improvement. At first glance it may seem that larger yearly profits would come if you had more than one breed, but the best way to judge such things is by the past. Such a view would reveal the fact that almost invariably the men who have attained full success, both financially and in the minds of their fellow breeders, have been those who have centered all their efforts upon one breed of sheep, cattle, horses, or what it may have been. If to some extent you want to guide your plans according to the experience of a majority of the best sheepmen, you certainly will never have more than one breed. That undescribable friendly feeling and united purpose which exist among breeders of all good breeds seem to be barred out and lost when two or three breeds place you in the "miscellaneous" class, and really you yourself hardly know what your purpose is. Your financial success in pure-bred sheep breeding will in the end be great just according to how much better your sheep have been than those sent out by others engaged in the same business. If your study is divided into two or three parts, does not that, first of all, weaken your ability to produce sheep of high type? Once that is decided there is no comparison in the remaining details because the successful breeder and advancer of one breed has a distinct satisfaction that no other class of breeders can enjoy. So it all sums down to what your real desire is. If you have no definite purpose, are not deeply interested in the advancement of the sheep industry, and do not value friendly relation with the best stockman of the nation, you may attain a so-called success producing several breeds, but if the opposite is your purpose and your desire is to be really successful as a pure-bred breeder, our suggestion would be "One breed and one only."

"CLOVER HILL" BREEDING EWES.

A "snap" on part of the young Shropshire breeding ewes at "Clover Hill Farm," 1909.

Founding a Pure-Bred Flock

When founding a pure-bred flock the main thing to consider is how to prepare to get the largest possible profits after you have been breeding. Sales from the flock will be to other breeders and farmers, therefore the important part is to produce that class of individuals and the breed for which there is widest and greatest demand. By studying your surrounding conditions, reading about and observing the experience of others with different breeds, and noticing the exhibits at leading fairs, you will readily see that the Shropshire is the universal breed. Its high position has been won by real merit and the good results it has given under all conditions. The Shropshire is the breed with which you can do the largest and most profitable business. However, there are many Shropshire flocks and the demand for really superior individuals is growing stronger and stronger, while there is a comparatively smaller demand for medium or low-class registered sheep. Those conditions point to the important fact that the wisest plan in founding a flock is to get such superior individuals that they will be far and away above the average and be admired by all who see them. Size is desirable and the more size you are able to obtain in the individuals of your foundation flock the better it will be for you. The way to raise big ones is to get big ewes—do not let anyone tell you that little ewes as a class will raise big lambs. For years the writers have bred for size, and the use of big ewes—not little ones—has brought us success. Purchasers, generally, demand the big ones and that fact is one to always keep fresh in your mind. If people like your foundation flock they will come to you in the future for rams. There are a few old reputable breeders who have always bred for size with quality, and ewes from one of those flocks would breed size for you. Steer shy of the many flocks which have been bred for quality, and quality only. It is sound policy to obtain top ewes from the most reputable flock possible—then your future flock will be larger and better and you will have greater demand. If your foundation comes from a rather ordinary flock what you would soon offer would not be any better than the average pure-bred. The object in view must be to raise superior sheep—superior because larger, stronger constitutioned, heavier fleeced, more correct in mutton conformation, and nearest to perfect Shropshire type. The good ones will always sell. Lay the foundation for your flock in such a man-

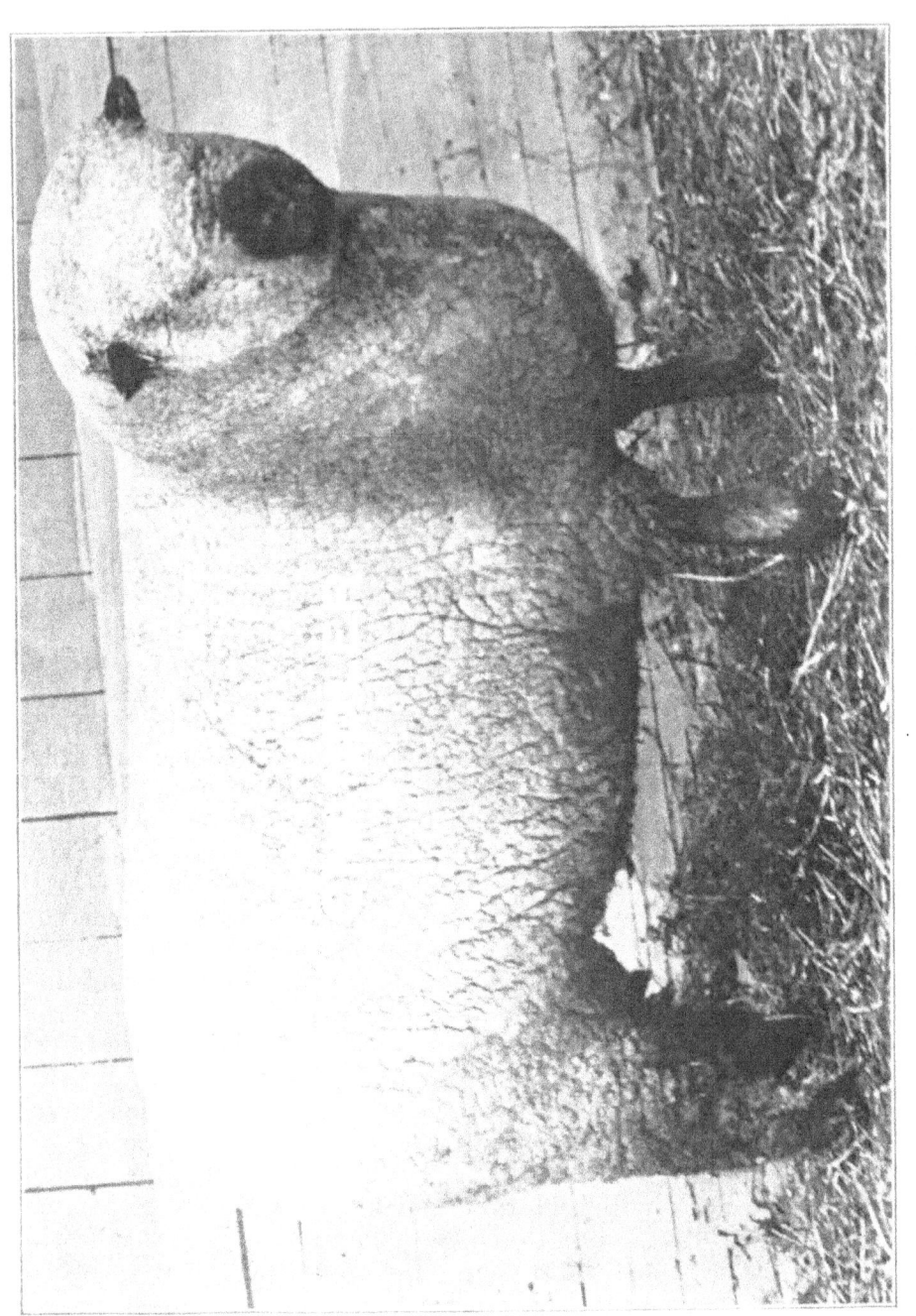

"CLOVER HILL'S GOLDEN DUCHESS" (298877)

First Prize yearling Shropshire; ewe and one of Champion Flock, Iowa and Minnesota State Fairs, also First Prize yearling ewe, Champion ewe over all *ages*, and one of Champion Flock *International Live Stock Show*, *Chicago*, *1909*. Owned and exhibited by Chandler Bros., "Clover Hill Farm," Chariton, Iowa.

ner that you will produce very few common ones. Observation and experience teach us that the first thing for a beginner to do is to build up a ewe flock of unquestionable merit for both individuality and breeding.

Get thoroughly fixed in your mind the characteristics of a large good mutton sheep, also true Shropshire type, then you are ready to visit some old-established reputable flock. The best breeders will be only too pleased to show you through the whole flock at any time and answer any questions you may wish to ask. Purchases from such an old flock will give better results than if you were to buy a few at different places and finish with a lot presenting three or four different types.

The general make-up of a sheep cannot be studied too closely, and it is well to have a system in your examination, because when purchasing a foundation flock you should carefully examine each individual rather than giving a simple glance over the bunch. Commence examination at the nostrils, which should be well extended. Nose, face, and whole head should be short and broad, neck short thick firm and smoothly blended to shoulder. With your fingers together the back can be examined for both straightness and covering of natural flesh. It should be straight from top of shoulder to loin and with as little droop as possible on to the tail-head. By natural flesh we mean the covering of lean meat on the bones. A well-covered back is firm and does not feel bare to the touch.

The loin should be wide and thick, the tail-head wide and well set up. A sheep's hind-end should be like two pears placed side by side because in that case there would be so much meat between a sheep's hind legs that they would have to be wide apart and the outside flesh covering would also be very good. It is well then to handle the back a second time because the sheep may be standing in a different position. Also notice the spring of rib because when the ribs come immediately and well out from back-bone it increases the width of back and gives more room for the internal organs. Then examine to find the depth and width of chest because a wide deep chest strengthens constitution by giving ample room for the heart, digestive organs, etc., to properly do their required work. To give good results either in the breeding pen or feed lot a sheep must have strong constitution, and narrow-chested, straight-ribbed sheep rarely, if ever, prove profitable. Straight strong legs are a necessity under breeding sheep and the pasterns must be strong and although this is more important with rams it affects ewes to the extent that they may drop rams for breeding purposes. When the sheep has been examined for mutton form, constitution, legs and feet, the fleece should be carefully looked through. A Shropshire fleece

21

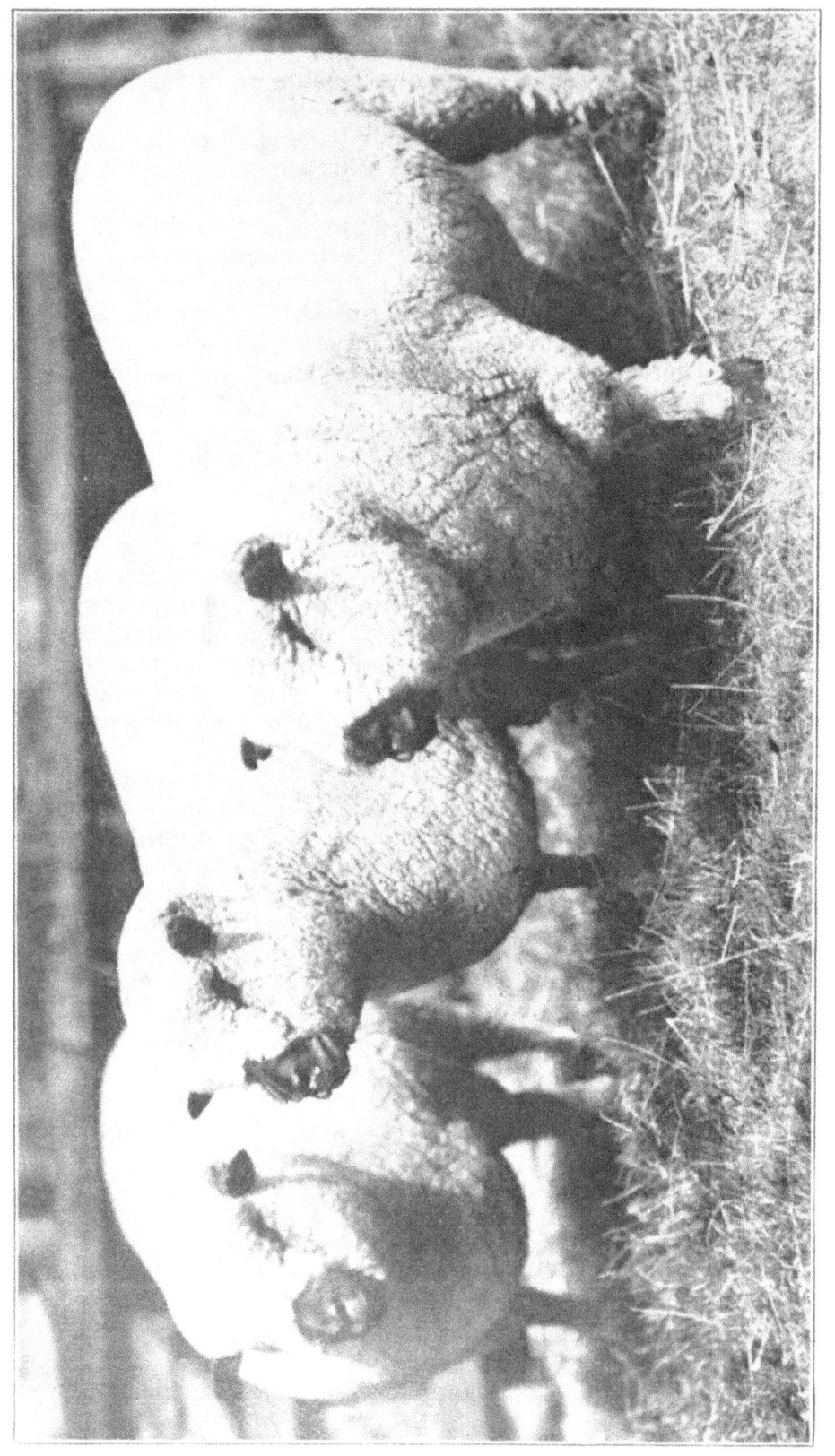

THREE "CLOVER HILL" YEARLING SHROPSHIRE EWES.

One of them "Clover Hill Aristocrat" A. S. A. 278958 won First Prize as yearling ewe and Champion ewe any age at Iowa, Minnesota and Missouri State Fairs, the St. Joseph Inter-State Fair, and the American Royal Show, 1908. Exhibited by Chandler Bros., "Clover Hill Farm," Chariton, Iowa.

is fine in texture, dense, bright, of good length, and with a uniform close crimp.

There should be a dense wool covering on the belly and inside the legs as well as on the body because this not only increases weight of fleece but it is protection from all troubles due to lying on wet ground. There should be no dark fibre in the fleece and as little as possible on top of head. Wooling should extend well down on legs, and in ewes strict femininity must be sought for.

It is well to place much stress upon the breeding and the advantages obtained by going to an old flock when making purchases. Pedigree is simply the record of an animal's ancestors, and a sheep whose sire, dam, grandsire, granddam, great grandsire, great granddam, etc., have been noted sheep with exceptional merit, will surely give better results in the breeding flock than one whose ancestors have been practically unheard of or unknown altogether. Successful breeding of livestock is a life work and the foundation must be laid slowly and well. By founding a flock with judicious selections from the best breeder the greatest factor is achieved—a sure and certain basis upon which to build up a flock of the highest class. Many breeders have never been able to eliminate the results of a bad start with the ewe foundation, despite the fact that they have used high-class rams for many years.

The right sort must be selected, even if an apparently high price has to be paid. We say apparently because the best specimens of the breed rarely prove to be dear in the long run while second-rate ewes are not cheap at any price and may be a constant source of disappointment. After having obtained the desired ewes, the first few years should be spent in improving the ewe flock.

To simply show the efforts that the writers have put forth to strengthen our own flocks and offer some to our customers from the most reputable English flocks we wish to here give a copy of the official report of the number of Shropshires exported from Great Britain:

	1908	1907	1906	1905	1904	1903	1902	1901	1900	1899	1898
North America (U. S. A., Canada & Newfoundland ..)	1774	1427	1057	253	217	65	424	364	481	677	314
South America....	205	569	657	497	275	288	48	104	205	282	438
Australasia	11	212	82	153	66	331	132	228	60	41	46
South Africa......	21	45	52	54	27	46	5	29	—	76	132
Russia, Germany, Spain, Sweden, etc.	35	61	54	75	118	58	84	265	118	10	62
Totals........	2046	2314	1902	1032	703	788	693	990	864	1086	992

23

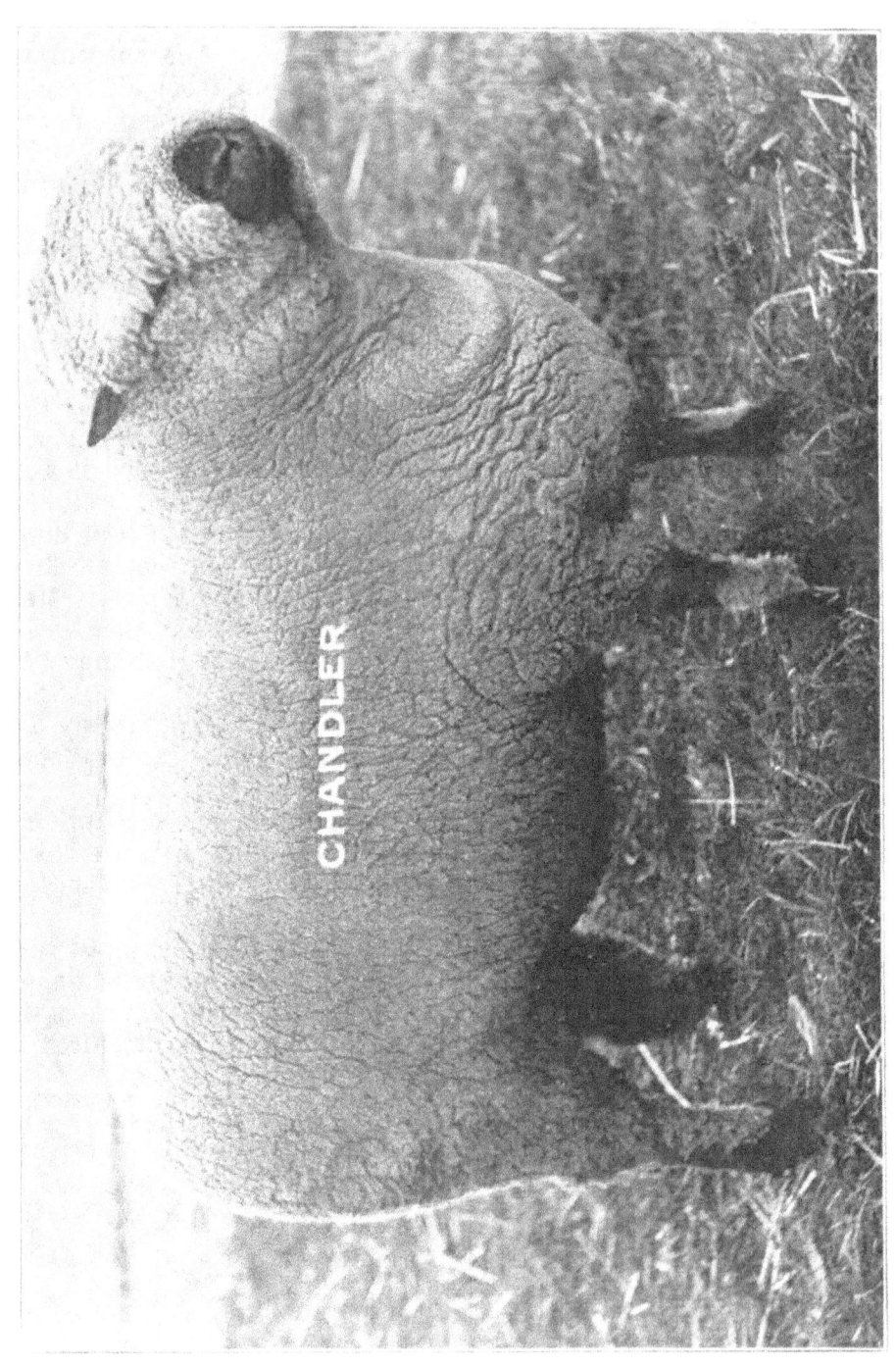

CHANDLER

"CLOVER HILL'S RIGHT STAMP" (298970)

A sire now in use in the "Clover Hill" Flock of Shropshires,

Of this 1774 Shropshires to the North American Continent in the year 1908, 1,117 head went to Chandler Bros., "Clover Hill Farm," Chariton, Iowa; 218 head to various points in Wisconsin, 121 to Canada, 106 to New York, 95 to Minnesota, and the remaining 117 head to various places. Cornbelt sheepmen can be especially proud of the above official report because, as is noticed, Iowa took more English Shropshires than all the rest of the world.

Selection of Sires

The ram is head of the flock and many call him more than half of it. In the ram, a breeder should get a noble specimen of what to his judgment is as near an ideal Shropshire as can possibly be obtained. Some say to select a ram strong where the ewes are weak—that sounds alright but you could continue such a practice your lifetime and still be see-sawing back and forth. One lamb would have the faults of the sire and another the faults of the dam. You are willing to pay a long price for a ram because you want him to stamp your ideal in the flock— so be sure the ram is right in every particular. After your foundation ewe flock has been carefully selected look to the ram for the rest. If the ram is near your ideal his lambs will be much like him, then get the next ram again as near your ideal as possible and keep right on that way—soon your flock would be pleasing you. If it were possible to get rams always just as we want them we would never look back at the ewe flock but would know that using good sire after good sire would bring the flock around absolutely right. But sometimes we have to do with some little fault in the ram so it is well to have a thorough knowledge of the ewe flock. But our suggestion would be never to sacrifice anything in the ram if you can avoid it. In no case let fashion lead you to forget that the ultimate object in breeding sheep is to produce mutton and wool at a minimum cost and with quality that will sell readily at top prices. When choosing a ram secure one with lots of size and decided masculine character, heavy bone and a bold wide walk, vigorous robust constitution, full chest development and well sprung ribs, wide loin, straight strong back and well filled hind-end, a broad wide head—the kind that anyone could tell was the head of the flock. We want a ram to be a ram all over and all the time. We use big rams to get

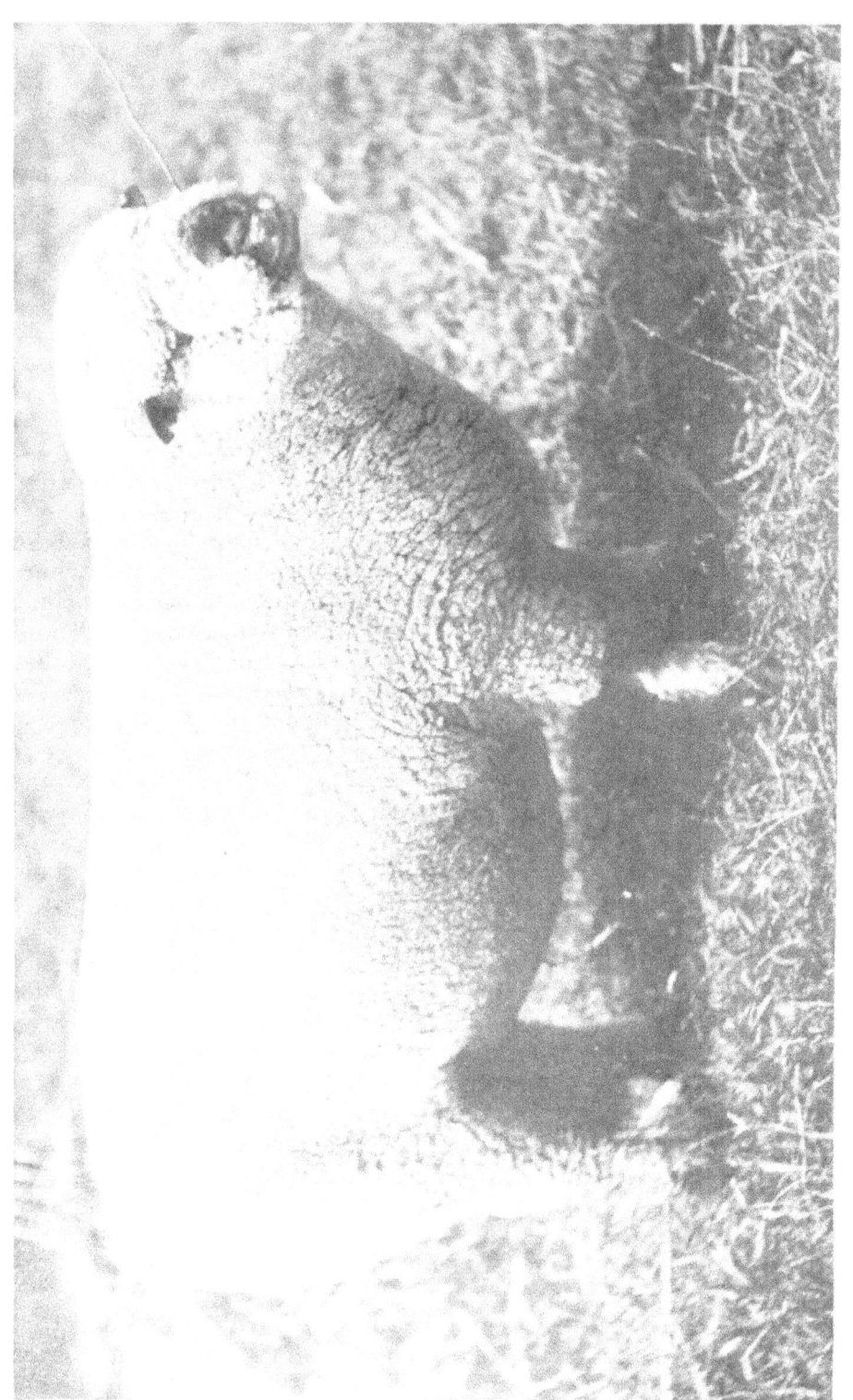

"CLOVER HILL'S ROYAL KNIGHT" A. S. A. 300970
A sire now in service in the "Clover Hill" flock of Shropshires.

big lambs, and heavy shearing rams to get good fleeced lambs. Some say "use a little ram"—we never chanced it but once and we got little lambs that never did grow big. No one ever made gold from brass and we do not believe that anyone can ever get rams which will mature at 200 to 250 lbs. in breeding condition from a small ram. We would not recommend such practice although it might be possible—we would not chance it again. If you can get a really good ram which is a big fellow and from big sire and dam, he is worth a whole bunch of the little good ones, and if his lambs shear 12 to 15 lbs. of clear white dense wool it will bring you more money than if there was only 8 or 10 lbs. of it.) Get the ram good and then as big as possible. Go to an old reputable flock that raises the big ones that are good. Be very careful about the breeding of the sire you purchase—do not let some unreputable breeder even give you one—and in this case understand the word "reputable" in a very strict meaning. Upon your fingers count the top breeders from whose flocks a ram would be reliable for the very best results, then select from those breeders what you can count possibly on the fingers of one hand and you would be getting down ready for business. When you get to breeding good ones there are very few places you can get the ram you really want. Pick out the one flock which you think is best of all—go and see the whole flock and if that breeder has not your ideal try the second best flock. The success of your flock depends largely upon the ram you put at the head of it, so "be sure you are right, then go ahead."

Influence of the Ram on Prolificacy

The question of the relative influence of the ram and ewe on prolificacy has been much debated, and rightly so, because it is a matter of great interest in breeding. There are two principal questions in this connection; one being as to whether the ram can create increased prolificacy in the ewe as the outcome of a single mating; the other being as to whether the male can transmit an increased tendency in the female progeny to the same because of inheritance. In the controversy some have claimed, and believe, that the ram exercises no influence on prolificacy, while others say that the ram does exercise an influence which is far-reaching. Some breeders go so far as to say that the influence thus exerted is as great as that exerted by the ewe, but a majority concede that the ram's influence is at least less than the ewe's.

Let us first endeavor to determine whether the ram does influence prolificacy in procreation; and, second, what is the relative strength of the influence compared with that of the ewe. With quadruped females capable of producing more than one at birth, and which may produce one, two, or three as the case may be, the different results are the outcome of some influence exerted on procreation in the dam rather than the result of chance. In the case of sheep, it is evident that such influence comes entirely from the dam or entirely from the ram, or from both, or it comes from one or more of the sources named, aided by external conditions such as food and environment. The ewe capable of bearing twins does not always produce twins, and why should there be such variation. That the ewe does exert an important influence on prolificacy is universally conceded, for while some ewes produce only one lamb at birth, others produce two, though mated to the same ram for successive years. It may be asked, then, does not this prove that the ram does not influence prolificacy? We answer no. The most that it can prove is that the ram does not exercise as much influence on prolificacy as the ewe, because the same ram mated with other ewes will in some instances result in but one at birth and in other instances in twins, which at least makes it possible that the ram does exert an influence on prolificacy.

It has been noticed that when but one ram is used in a flock, the proportion of twins from the earlier births is greater than from those later. From this it has been argued that this result follows from the greater vigor possessed by the ram. But if vigor in the ram influences prolificacy then, by parity of reasoning, it does in the ewe, hence it is conceivable the result stated may come partly, or chiefly, or even wholly from the ewe, as the more vigorous among the ewes come first in heat. But it is almost certain that some of the influence resulting in plural births comes from the ram, as, if the said ram, enfeebled by excess of service, was then turned in to mate with ewes of another flock, equal in prolificacy and vigor to the former, it is almost certain that a less proportion of twins would be produced in the earlier births.

From the facts stated, therefore, it would be correct to say that it is extremely probable that the male does exercise an influence on prolificacy. But thus far, only vigor has been recognized as the source of such increased prolificacy. To this may be added judicious feeding. The ewe pastured on rape or second growth clover for some time before mating will be more prolific than the ewe confined on dry prairie grasses. It is taken for granted, then, that vigor and food do exercise an influence on prolificacy, and it is almost certain that these influences are so operative through the male as well as through the female. Once grant that the ram does exert such influence as the outcome of judicious feeding, and it is then not difficult to show that this influence on the part of the male will be strengthened or weakened, as the case may be, by inheritance. Some persons claim that the ewe influences only certain parts of the organization and that transmission in the ram influences certain other parts.

That we do not exactly believe. It has never been proved and until it is, the conclusion is justifiable that the influence of both parents extends to every feature of the organization, including breeding tendencies, not necessarily in equal degrees nor always in the same degree. Then it follows that the power to transmit tendencies to prolificacy, or the opposite, inheres in both male and female as the result of inheritance. No one will doubt this in the case of the female, but many do doubt it in the case of the male. If it is true that the ram does exert influence on every part of the organization, including capabilities in the line of performance as well as physical features, it remains a fact that the ram, as a result of inheritance, does transmit tendencies to increased prolificacy, the influences that govern reproduction being so affected by the degree of this inheritance, but this may to some extent be modified by the influences of quality in foods and by vigor inherited or acquired. To increase prolificacy in ewes, therefore, it would be in order to choose rams from ancestry that have produced twins for generations previously.

From what has been said, it will be obvious that though rams should have the power to transmit tendencies to prolificacies as well as to beget prolificacy in the female, it will be impossible to determine the degree of the influence which they will thus exert, absolutely or relatively, since, as previously intimated, it will be a varying quantity because of the influence from the various sources mentioned, but it is to be expected that the influence on increased prolificacy will be greatest when the influence exerted by the male and female operate in conjunction rather than in opposition. In other words, when both ram and ewe come of ancestry noted for prolificacy, than when such inheritance belongs only to one parent. When making purchases or selecting from your young stock for your breeding flock, it is well to keep these facts in mind and select those from twin-producing families. Then by judicious feeding, a large percentage of the flock can be caused to produce twins, and that is very desirable.

Value of United Effort

United effort among sheepmen is not usually valued as highly as it should be, and breeders on the whole would receive much greater returns from their flocks if such were not the case. No one, regardless of how high-class his flock may be, can achieve as great success alone as he can by working together with neighbor breeders. We do not mean that they should be partners from a financial standpoint, but all should be partners in helping advance each others' flocks and the industry in general. There must be no such thing as jealousy, in fact, quite the opposite should be the case. Breeders should rejoice in the success of their friends, and apart from that right feeling, anyone's success helps all others who are engaged in the same business. By free conversation over all matters pertaining to the flock, neighbors of different vicinities can get greater results. The old saying "In unity there is strength" applies to the sheep industry the same as to others. Especially should producers of the same breed work harmoniously together to advance their cause for the good of all. How great would any certain breed become if all the sheep breeders in the world united for it? The good results would nearly be beyond our imagination, but even in some of our present single breeds we believe there could be improvement in unity of breeders which would be as great in comparison with existing conditions. Some good breeds of live-stock have almost been crushed out of existence by the fact that many of the breeders took an "independent" standing, while closely united efforts of all have championed many others. It seems to us that this unity of effort among sheepmen should be improved as much as the flocks, and the results from such action are surely so promising that no one should refuse to do their part. Not only does the complete unity make a breed more famous but when breeders of a certain vicinity work together they create a reputation for themselves. The true breeder is he who is glad to lend every possible assistance to beginners and the latter are indeed very fortunate when they have such help and advice. There are so many things that two or three sheepmen can do that one alone sometimes hardly cares to, and among these perhaps the most important is the purchase of a sire which really meets their ideal and needs of the flock. Especially is this true with pure-bred breeders who have advanced their flocks to a high standard, because the class of rams to then meet the requirements are very scarce and hard to obtain. The ram may be found and while one breeder would feel hardly able to make the purchase he would gladly do so in partnership with a neighbor. Some of the best sires in the world have been purchased in this manner and have made great improvement in two or more flocks at a minimum cost to the owners. Even in the purchase of foundation ewes, it is ofttimes the case that one farmer hardly wants a large enough number to make him feel justified in going to the expense of visiting a leading and reputable flock and shipping home his rather limited purchase, while if his neighbor also wanted an equal number, one of them could make the trip and general expense would be cut in two. That is where unity helps many who otherwise would not venture in the beginning. Even when producing sheep for mutton market, farmers are just as well repaid for working together. In buying their rams at once they get lowest price, but one of the greatest advantages for the farmers who have a small number of sheep is that they can unite for a car-load when shipping lambs and thereby get largest income from the flock. Wool can be put together to attract best buyers, and in all respects it pays to unite. Exchanging ideas with each other for united effort along the same line will enable breeders of vicinities and nations to achieve success which otherwise is impossible.

"CLOVER HILL'S JEWEL" (299970)

First Prize ewe lamb at Iowa and Minnesota State Fairs, Champion ewe over all ages Iowa State Fair, and one of the Champion Shropshire Flock at International Show, Chicago, 1909. Owned and exhibited by Chandler Bros., "Clover Hill Farm," Chariton, Iowa.

Explanation

Some photos in this book were used before we obtained copyrights from the United States Government and we learn that they have been copied by other sheepmen to use on letter-heads, envelopes, etc.; therefore, all photos recently taken have C H A N D L E R worded across them, and we have also obtained full copyright for protection so infringers can be prosecuted. Many years of hard work and careful breeding are required to produce such superior Shropshires and the photos of the individuals which are the result of our work should not be used by others— it would be injustice to us and misleading to the public.

Another way that breeders of inferior or medium-quality pure-bred flocks try to mislead those interested in Shropshires is by saying "won all prizes shown for," meaning to convey that the flock has been winning at State or National Fairs, while in reality they were shown at some little County Fair with little or no competition. Other breeders further along will win flock or some prize in restricted classes at a State Fair and say "First Prize Flock at ——— State Fair," thus intending to lead breeders to believe that they won Champion Flock over all competitors. Even some big exhibitors having two, three or more breeds, one being the Shropshire, will say "won more prizes than all competitors." but the actual facts probably are that there was very little competition in the other breeds and in Shropshire Open Classes they won very little. Some breeders would go a long way to turn a round corner rather than turn a square one. People interested in Shropshire facts, and especially intending purchasers, should closely study the different flocks.

Every prize mentioned under photos of sheep in this book were won in "open to the world" competition unless stated otherwise.

Little Facts for Both Beginners and Breeders

Energy is what wins.

Size with quality is the best policy.

Aim high for a definite object.

Success does not come, you have to go after it.

Superiority in both type and individuality of your flock is the best trade-mark.

A clean record is the greatest kind of success.

The period of gestation in sheep ranges usually from 21 weeks and 5 days to 22 weeks.

Most of the things that breeders attribute to misfortune are due to ignorance.

It is a disgrace not to do one's level best to succeed.

Any breed is worth caring for well, but some give larger returns than others.

First-class sheep do not come from any but first-class firms.

Intense earnestness, perseverance, and familiarity with the minutest details of the sheep industry are the chief elements of success.

Placing grain in "self-feeders" is far from being the most economical method of feeding.

To please just one customer will give you an opportunity to please his friends and acquaintances.

Wisdom consists of doing the things it would be foolish not to do. Every farmer is wise in raising sheep, and good ones, too.

Plan your work, then work your plan. The breeder who doesn't will never reach a high standing.

Foundation breeding material from old, reputable flocks can be relied upon to give uniform results of the highest class.

Salt should be kept before the flock, and although it does not affect digestion, it tends to increase consumption of food and improve nutrition.

There is always a demand for scarce articles, therefore to obtain the highest prices you must produce sheep with such quality that they will be hard to equal.

It is not necessary that a man know all things, but in order to succeed he must know who knows that which he does not know, and go to him for it.

Anyone will work hard when all is coming his way, but the sheep breeder who climbs to the top of the ladder of success is he who makes stepping stones of what are stumbling blocks for others.

A breeder's reputation is based upon the class of sheep he produces—the enviable one being created by the superiority in sheep sent out.

And if you fall—why rise again! Get up and go on; you may be sorely bruised with your fall, but is that any reason for lying still, and giving up the struggle cowardly?

Sheep are creatures of habit and should always be handled by the same regular and quiet attendant. Dogs and strangers should be kept from the feeding pens at all times if possible.

In the pure-bred business, a breeder's friends might nearly be termed an asset. Friends are our mirrors and should be clearer than crystal.

It is highly profitable to correct a mistake. If you resolve to do so, the care you exercise in avoiding them will give you a less number to correct.

Large oaks from little acorns grow, and the big breeders have been little some day. This is satisfaction to the beginner because what has been done can be done again.

The business which is conducted on the basis of a hope for permanency must give value received or it will die.

Old sayings are the best or they would long ago have been forgotten, so there must be great truth in that familiar one "blood will tell." The scrub ram gives undesirable results which are a detriment to any flock and lower the profits derived therefrom.

"Opportunity knocks once at every man's door." When it comes to you in the pure-bred sheep business, be sure you are awake to answer the bell, because if you don't some one else will and then gain the trade of your vicinity. Successful breeders are "live-wires" all the time.

Make your methods of breeding an object lesson for improvement to all who may visit your flock.

Breeders who are another stone in the solid foundation of the Shropshire are those who in an unselfish manner do all possible good for the breed and others interested as well as for their personal present gain.

If there are County Fairs in your vicinity and some breeders are not exhibiting because the prizes are small, you should see the Board of Directors to have the premiums increased. They will no doubt give a liberal classification if good exhibits are assured. It is very essential to the greatest success of the Shropshire breed that they are well represented at the small fairs throughout the country, because many beginners get their first inspiration and impression there.

There are two classes of men who have Shropshires, one including those who have Shropshires with other breeds simply for the money there is in it—they are not real Shropshire

"CLOVER HILL'S 6501" A. S. A. 300025

First Prize yearling ewe and Champion ewe any age in Iowa and American bred classes, Iowa State Fair, 1909, bred and owned by exhibitors Chandler Bros., "Clover Hill Farm," Chariton, Iowa.   Also one of the Champion Iowa-bred Shropshire Flock, 1909

men, although they may claim to be. The other class includes those who have Shropshires exclusively and have a deep interest in the welfare of the breed and all who found Shropshire flocks. Support should be given to the latter class because their services are superior and it requires strong united effort to insure continual great success of any breed.

Success as a breeder is gained by many years of careful selection and breeding for size, type, character, mutton conformation, and dense fleeces. When purchases of breeding ewes or rams are made they are selected from the most reputable flock within reach because individuals from only such a flock will give the desired unvarying results. The "old war horses" are those who have looked to the future and built for it. Do not expect success in one or two years.

Standards which determine economy in the purchase of breeding sheep do not lie in dollars alone. Over against them must be considered the really important measure and that is value. An expenditure, no matter how small, is extravagance if it brings no returns. If you purchase rams or ewes simply because they are low-priced, and they give very little or no returns it is extravagance. It does not matter how high-priced a sheep is, just so he is good value.

What the Shropshire Has Done for the American Farmer

The Shropshire sheep enjoys the distinction of having been the solid rock upon which the foundation was first really begun to make America a mutton-producing country. It seems nearly as if Providence had piloted the breed to this country for improving or "opening up," the same as Columbus came first to a country which later grew and improved far beyond their greatest ideas. But it was a country which, if left untouched, would still have been a wilderness in comparison with what it is to-day, but a new country was needed by the people of the world and it was discovered. When the Shropshire sheep was first introduced into America there were practically no mutton sheep, but they were needed, and they came. There was a place for them, and had they not been introduced the agricultural population of this country would be at a great loss. What the Shropshire has done to the sheep industry is nearly too vast to comprehend. When it is considered how the favorable results which came from the first Shropshires

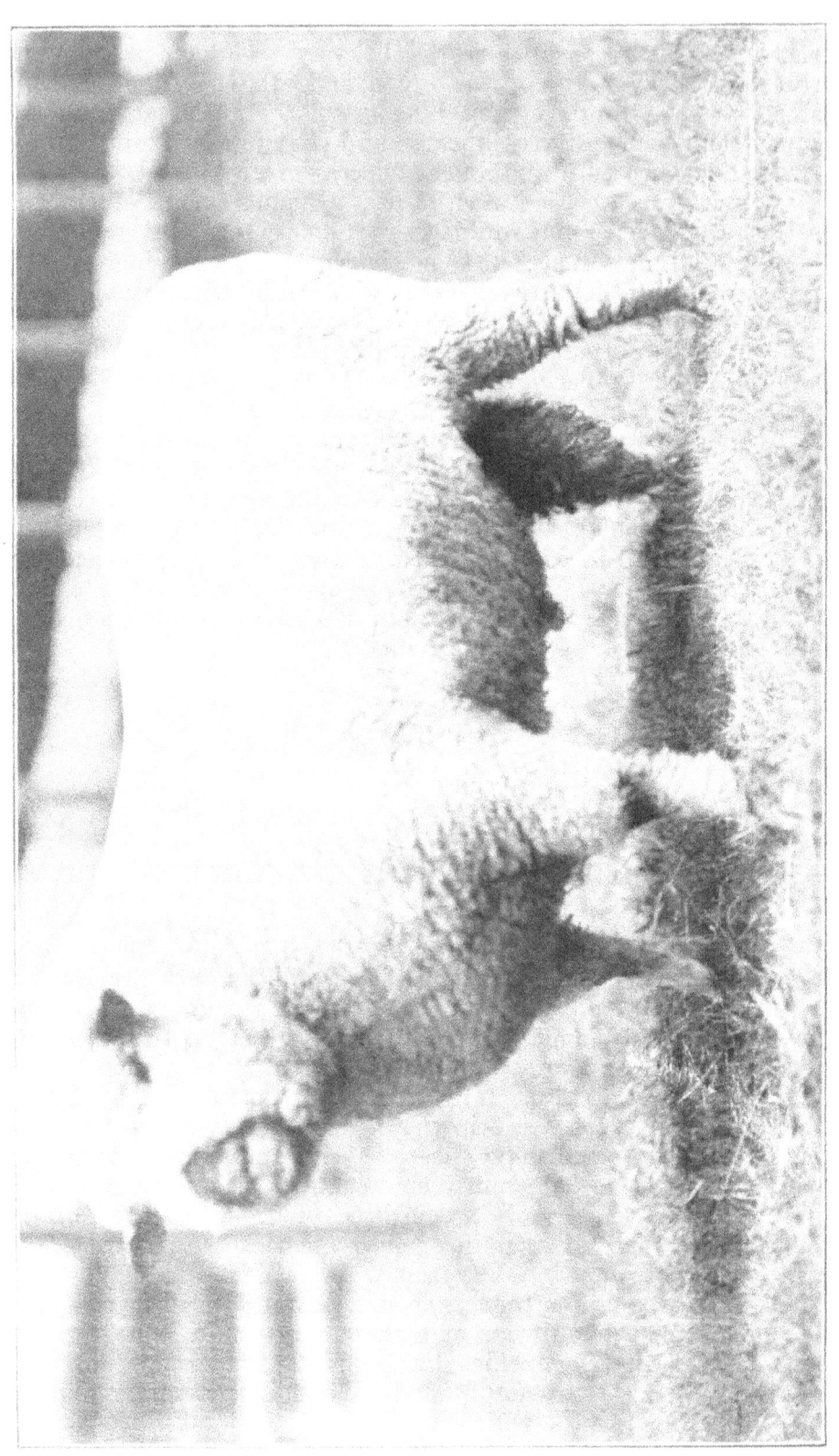

THE SHROPSHIRE SENSATION OF 1908.

"Clover Hill Emblem" A. S. A. 75945, a two-year old exhibited by Chandler Bros., "Clover Hill Farm," Chariton, Iowa and a Champion over all ages at nearly all State and National Fairs. The Shepherd's Journal of Chicago stated editorially "this famous two-year old Shropshire ram has proved himself an invincible Champion. It is safe to say that a more typical ram of the breed never has been seen in our show-yards." Used as a sire at "Clover Hill."

were a stimulant for expansion of mutton sheep production, it must be granted that the Shropshire is a breed of great merit. They have proved profitable from the very beginning and the present condition of the mutton industry traces to a greater or less extent back to this one breed. Of course, to-day there are many mutton breeds, but they have come along the path previously paved by the Shropshire. It came and made clear the fact that mutton sheep were required. Their strong constitution made them do well under all farm and climatic conditions, their mutton was of such quality that it filled a market requirement which had never before been met, their fleece was of good weight and with special density under the body to protect the sheep when lying on damp ground, and not only did the pure Shropshire fulfill the requirements but they strongly impressed these qualities in their offspring when crossed on other sheep. When people began to learn of this the demand for Shropshire blood increased and has steadily increased ever since. The breed greatly improved the common sheep and made an excellent cross with the fine-wools, and not only did the sheep produced by such cross-breeding meet the requirements, but they have broadened the mutton demand into all sections of the country.

First, the Shropshires came to one section and the improvement was not far reaching, but it soon began to spread and has continued up to the present day. The comparatively limited number of rams available in early years even retarded the improvement that sheepmen wanted to make when it was generally known what advancement mutton sheep were making. As better mutton has been produced from year to year the demand has not only increased but has steadily changed for the best class. As the country's population has learned that first-class mutton is obtainable a much greater quantity is being consumed. These facts are especially verified by the great change in market conditions. In days gone by there was not much discrimination when a car of sheep was sent to market, but now the price varies exactly according to the quality of sheep being offered. Good ones bring high prices and the undesirable sheep bring quite low prices. Has the Shropshire not been a main factor in bringing all this about? They proved that Shropshire-cross mutton was good mutton and thousands of people were ready to buy that class of meat. As the sheepmen have learned that is what is required, the breeding of Shropshire sheep has expanded but that expansion has hardly been as great as is the demand. Farmers have been well pleased because the Shropshire flock has taken a place on the general farm that nothing else seems to fill. The fact that the Shropshire is at the present day raised in every State in the Union is because they have given results which are sought for

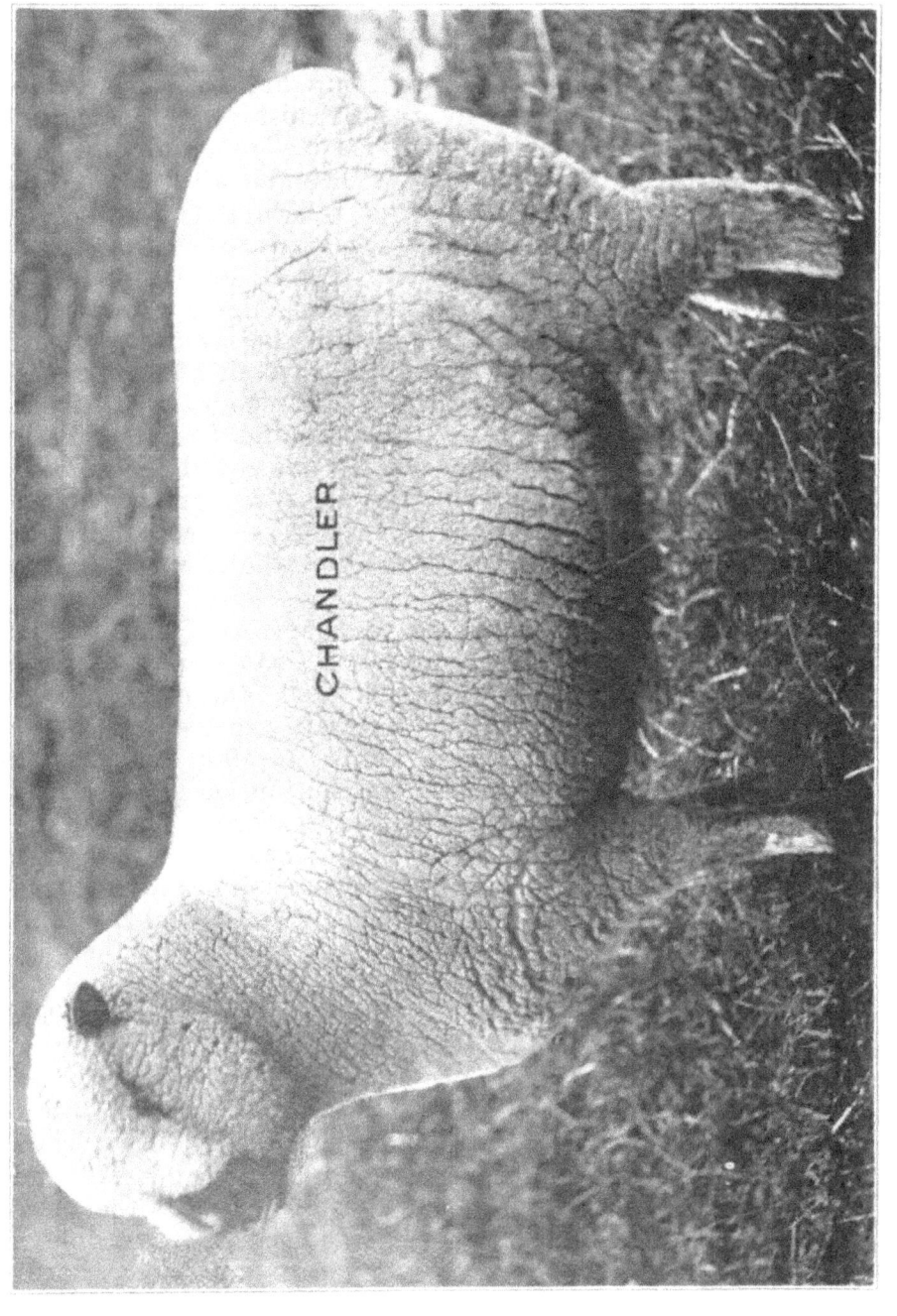

"CLOVER HILL'S 5369 A. S. A. 28503.

First Prize Shropshire ewe, two years old or over, Iowa and Minnesota State Fairs, the St. Joseph Inter-State Fair. Champion Shropshire ewe any age Minnesota State Fair. Also one of the Champion Shropshire Flock at Fairs mentioned. Exhibited by Chandler Bros., "Clover Hill Farm," Chariton, Ia.

by the agricultural population and they have the mutton that is sought for by the meat-eating public. They are the universal sheep because they have "made good" from the beginning. Should not all sheepmen be thankful for this because it has placed the sheep industry of America on a solid foundation and made it a business of stability? Had this breed never been introduced here, what would have been the present condition of the industry? An answer would have to be based on supposition, but it is safe to say that had the breeders started with sheep which were not so hardy under all conditions and did not produce such a good mutton carcass when crossed on other breeds, their interest would have been slackened and desire to expand would have been cut off. Also, if the public had not been continually getting a better grade of mutton their demand would have contracted instead of expanding. But the Shropshire pleased the breeder and pleased the consumer, and our country is thankful for such a breed. The great strides forward which have been made are a benefit to all and we do not believe that anyone does not really in their own mind give all due credit to the Shropshire. The greater portion of America must be a mutton producer, and none other than a mutton sheep of the highest class fills the bill. The Shropshire has done it, is doing it, and always will. They are the sheep for the farmer and there are good reasons for it, so many that none but those who raise Shropshires can ever really learn them all. Constitution has been a most desirable characteristic and farmers want such a sheep that doesn't need to be kept in a hot-house and that will not suffer if caught out in a storm. Not only are they naturally strong but their dense fleece, which completely covers the body, is the greatest sort of protection. Probably they are in a class by themselves when it comes to raising twin lambs. Some breeders may say that they would rather have one strong lamb than two weak ones. Yes, but wouldn't you rather have two strong lambs than just one? That is what the well-cared for Shropshire ewe will bring. Only a short time ago we saw two Shropshire ewes with seven big, lusty lambs on an Ohio farm. Of course that is an exceptional instance, but the fact that the two ewes gave birth to the seven lambs and were rearing them well only shows one of the breed's strong characteristics. That the quality of Shropshire mutton has been a principal factor in building up the American mutton business speaks strongly enough for itself, and the most desirable feature is that these good qualities come out very distinctly in the offspring when low-quality flocks have had Shropshire blood introduced. That fact has been a boon to American agriculture because the importance of a flock on every farm is getting more and more evident each year and had not the Shropshire given such good, all-around

results in the past the probabilities are that the sheep industry would shine under a different light than it does to-day.
To a certain point, the more Shropshire blood that has been
introduced into the flocks of a community the greater has been
the success of those breeders because they have met the demand for first-class mutton, and mutton is what the world is
calling for. The demand which the Shropshire breed has
created, and the desire of farmers to raise them because they
are so profitable, when coupled together is what makes the
Shropshire the exceptional breed that it is. Grade sheep producers have learned that it is the breed for them because of
the market demand, and pure-bred breeders achieve success
because the demand is keen for the Shropshire breeding material they offer for sale. Facts of conditions show that they are
in a class by themselves as a utility sheep for the general
farmer and in addition to that they are most beautiful for
those who love to have good live-stock as well as to have the
kind that are a financial success. Therefore, today the solid
old Shropshire is to be recommended to those who want a general farm flock that will raise lambs to bring the highest price;
and also to those who are engaging in a pure-bred business to
derive satisfaction, pleasure, and profit from it. The Shropshire has done well by its breeders and will do even better in
the future because it has built up its own foundation that has
stood the storms and is not floating on borrowed reputation.
It lives on facts, and facts are what count. The future must be
judged more or less by the past, and that is what makes the
Shropshire enjoy its present distinction.

Have an Ideal in Breeding

The word "Ideal" is perhaps the greatest in the vocabu-
lary of the sheep breeder or any other business man. It means
an imaginary standard of excellence, a plan or purpose of
action. One need only ask themselves, what good would ever
be accomplished or advancement made if man had not in
mind some standard of excellence to reach or some definite
purpose of action, to partially understand the great meaning
of the word and the important place it must occupy in our
life and work. The past success of different breeders has
been the result of their untiring effort to reach a certain ideal.
Ideals may vary, so also the results, but the one special factor
for the breeder to remember and study is to have a definite

purpose. If success comes without it, it is an accident and accidents seldom occur. To the breeder who has a fair start in the sheep business, the main consideration now comes before you, "what is your ideal—what goal do you want to reach?" At the very beginning a person cannot give to himself a satisfactory answer to that question, but after having raised sheep for a while and experience has taught a few lessons then the collected facts enable one to make plans for definite procedure in the future. However, this concluding of what your ideal exactly is should not be put off too long because the habit of being idealless soon affects you until there is no desire to accomplish any definite success. How many failures or "drones" are seen in this world which are the result of their having permitted surrounding circumstances to guide them? The number is certainly large but we simply mention them here for comparison and not to be copied after. On the other hand you hear people say of someone: "Oh! luck always comes his way and he makes a success of anything he puts his hand to." The facts usually are that such men have an ideal and are putting forth every effort along well planned and guided lines; they have a purpose of action—an imaginary standard of excellence. To have this ideal is all-important because there is a great deal of difference between the "bull-headed" man and the "thinker." They may possess an equal amount of energy and perserverance but the "thinker" has a purpose of action and his work is carefully guided by deep thought, while the other class hardly know what they do want. Of course, different conditions change results but in a majority of cases the failure to succeed in sheep breeding is due to the lack of an ideal. On the other hand the good results leading to great success have come because someone was working hard along a well laid plan to reach a certain end. As soon as conditions are right, be sure to have an ideal, then work incessantly along that line. You make the circumstances instead of letting them guide you. When things come up apparently to work against you, go straight through, or around if necessary, but never say "I can't." There are many stories about that "Can't" which are familiar to us all, but it must be an unknown word to the breeder who desires to be a real flock-master. No matter whether you are producing mutton lambs for the market or pure-breds for even the show-ring, success rarely comes except as the reward of a definite purpose which has been carried out by the breeder. "Hope" is the foundation for your having a flock—you have hoped to do a certain thing. Perhaps you have engaged in pure-bred breeding and now you hope to sell the kind of rams that will improve the flocks of your customers or possibly have an even greater hope to produce sheep which will win the greatest prizes offered at live-stock exhibitions. We live on Hope—hoping that such and such things will be after awhile. That is a good thing, but what is the real use of Hope unless we have an Ideal to work toward so our hopes will be realized? If your efforts are centered upon one definite object and are combined with perseverance, some time sooner or later your hope will surely become real. You may work hard every day of your life, but without an ideal prove a failure. The work guided to do a certain thing and continued until it is done is what crowns breeders with success. First they view surrounding conditions, then form their ideal—an imaginary standard of excellence, and set out and do not stop until it is realized. You must surely plan to do a certain thing before it ever will be done. If you want to sell high-class rams, that must be your purpose and then continually strive along well planned lines until you do offer them. Continually make changes in your breeding flock and methods of management until the desired sort are the result. Know what you want and do not stop until you get it. The highest desire of a pure-bred breeder is, of course, to win in the show-ring. It is not getting the premium money that gives satisfaction but it is the great honor of having brought forth sheep nearer perfect than anyone else has. That should be the highest ideal of those who found a pure-bred flock—say to yourself, "some day my sheep will win in the greatest show-rings." Having that object in view brings the other good things ahead. The winning sheep have to be more perfect in constitution, mutton form, fleece, type, etc., than any competitor's and when you are producing that class of sheep your sales will surely be good enough and you will have the extreme satisfaction of knowing that your sheep are helping to advance general sheep interests as well as your own. Men with such ideals are the ones that have made our pure-bred sheep what they are, and breeders with such a purpose are required for the future welfare of the industry. We were put in this world for some purpose and if our lot is sheep-raising, let's go at it with all our might. Some adversities will come the same as clouds occasionally overshadow the sun, but let us be sure to have our sun shining brightly as soon as the clouds disappear. If something happens wrong at home, overcome it and try to make things better for the future, and if someone wins over you in the show-ring, study to learn how he was able to produce a more perfect sheep and then try to make yours better. That kind of an Ideal will bring success to you, both financially and for the satisfaction and pleasure.

Selecting for Exhibition

At the end of October, or early in November, some six or eight ram lambs, and about the same number of ewe lambs (if required) should be selected, with a view to giving them a little more care and attention than the rest. This requires a great amount of care and discrimination on the part of the breeder. In the first place, we would recommend that double the required number be drawn out, the lambs from ewes which have previously bred exceptionally well, being specially noted. After this they must all undergo a close scrutiny and those that have defects which would debar their winning must be discarded at once, as it would only be time thrown away to prepare and feed for show an animal that could not in some degree be successful. What is a defect which cannot be passed over in a show-ring must be left to the judgment of the breeder.

The young sheep intended for exhibition should be allowed to run out daily until about April on an old turf for preference and at night should be housed. But should the winter prove very mild they are really better not housed at all until later. They should also receive a liberal allowance of what succulent food the farm will allow—such as turnips, kale, mangels and cabbage, ½ lb. to 1 lb. of artificial food, consisting of oats, linseed cake, and bran, and as much good clover hay as they will consume. Neither corn, beans, peas, or other highly nitrogenous food are considered desirable, as they heat the body and tend to derange the system, and if given in excess, the results will soon be seen in ricketty legs and a shrunken appearance of the barrel. Sheep grow and thrive better upon plenty of green food and good clover hay with a moderate allowance of grain, than with corn as the principal ration and a short supply of nature's own food. Many breeders have had a good pen of ewes or a valuable ram completely ruined by the injudicious use of too much trough food. A number of young or inexperienced shepherds make this great mistake and the watchful eye should be ever on the alert to counteract the tendency to use artificials too freely. The most successful feeders of show sheep are invariably warm advocates for a variety of green food, using, comparatively speaking, little else.

As soon as warm weather comes, the sheep intended for exhibition should be shorn. On no account wash the sheep intended for exhibition before shearing, as it may at such an early period cause death by chill.

"CLOVER HILL'S 6514" A. S. A. 300191

First Prize Shropshire ewe lamb, Iowa and American-bred classes, Iowa State Fair 1909, bred and owned by exhibitors, Chandler Bros., "Clover Hill Farm," Chariton, Iowa. Also one of the Champion Iowa-bred Shropshire Flock, 1909.

The sheep should, after shearing, have their feet carefully looked to and pared, and again undergo a close examination, with a view to finding if there are any that are too faulty for show purposes. If they all pass muster, they must be divided into small lots and pushed on (with judgment) as rapidly as possible as the summer shows will soon be here. Not more than three of the best rams should be put together—they are still better alone—and the ewes say in two lots. Sheep thrive and do better in small lots and in the case of rams it is absolutely essential to divide them as much as circumstances will allow.

Rams are not so likely to fight in lots of three as when two are penned together but the best animals should, if possible, have pens to themselves.

The sheep shed should be on a dry spot and a good system of drainage is essential. In practice it will not be found advisable to put the show ewes and rams in the same shed, and generally another building may be so arranged as to accommodate the ewes.

Exercise is, however, essential even after the sheep have been shorn and housed, and the chances of success are greater if the sheep have exercise daily. It develops muscle and strength, keeps them well on their legs, gives them an increased appetite, and promotes health, all which results in a thrifty growth and firmness of flesh. Exercising the show animals is pleasant occupation and if you are accompanied occasionally by a friend, so much the better, as the animals get used to strangers and become docile and quiet. Most successful men play and faddle with their sheep and notice their daily growth, and if they are not doing as they should, change the diet somewhat.

Preparing Lambs for Show

These should be chosen with care, examining each lamb minutely as to wool, skin, and other points characteristic of the breed, and as a rule early well-grown lambs have much the best chance. Let the lambs selected be dipped at once, repeating the operation a month or six weeks later. Give them as much room as possible on land which has not been fed with sheep, housing them at night, and supplying them with green food, and about ½ lb. of linseed cake, oats, and bran mixed (no corn). As the summer advances, house in the heat and glare of the day, and allow them recourse to a field for a few hours in the evening. As the show-time approaches, it may be advisable to trough feed entirely, and only allow them out for exercise, as they are apt to fall away when from home if not so treated, and unaccustomed to confinement.

"CLOVER HILL'S CAPABLE QUEEN," A. S. A. 278963.

First Prize ewe lamb at Iowa, Minnesota, Wisconsin and Illinois State Fairs, the St. Joseph Inter-State Fair, the American Royal Show, and Champion Shropshire ewe any age at Illinois State Fair, 1908.   Exhibited by Chandler Bros., "Clover Hill Farm," Chariton, Iowa.

Green Feeds for Summer

Rape

Rape is the most universal green plant grown for summer use with lambs and exhibition sheep. Perhaps this is due to the fact that it is easily raised and grows an abundance of feed on a small area, as much as fifteen tons having been obtained from a single cutting per acre. Dwarf Essex is the best variety and grows up very quickly being large enough to cut for use within seventy to ninety days after planting. The seed is about the same in size as that of the turnip and when sown broadcast it will require four to six pounds per acre. It can be sown in the spring as soon as warm weather comes and when sown in drills it will not require over three or four pounds per acre, yet it need not be spared because it only costs from five to eight cents per pound. Well prepared clean ground raises the heaviest and best crop. If not cut too short the first time there will be a luxuriant second growth which comes into good use for the field lambs in the autumn. The housed sheep do wonderfully well with rape as their green feed.

Kale

Thousand-headed Kale is another very desirable green crop and in appearance is about between rape and cabbage. Of course the leaves of rape are wide out while those of cabbage form a head, but kale would not come under either description. Possibly it is harder to raise than rape but it contains more dry matter. The best crop of kale comes from having drilled it in rows about 18 inches apart. Then when it has come well up it can be thinned in the rows according to the growth of plant that is desired. They should be left fairly close together or they will become so large that the stems will be woody. If it is your first year with these plants only a small patch of kale should be put out and then it can be seen how kale and rape compare.

Cabbage

As neither Kale nor Rape will keep in good condition very long after cutting, the exhibitor must raise some crop that will afford the necessary green stuff while away at the shows. Cabbage fills this place very well. Seed of the early varieties can be sown in a "bed" at the house and the plants transplanted as soon as spring opens up. This method is necessary in order that you may be sure of some fair-sized heads when you first need them. For the later cabbage some flockmasters prefer to drill the seed in the field but we have never had very good success in that way so have adopted the same manner of transplanting that we have with the early varieties. If you get a good cabbage crop you have assurance of keeping your flock in form throughout the fair circuit.

Turnips

The flock while away do all the better if they have some additional feed other than cabbage and their grain and as it should be of a succulent nature the turnip comes into place. In the early spring a small patch of turnips should be planted and although some prefer the Purple Top we have always obtained the best results with the White Globe. It doesn't matter so much what variety it is, so they are put in early and on a clean patch of soil. They should be drilled in rows so they could be well cared for and their growth stimulated. Mangels and Ruta-Bagas cannot be successfully used in place of the turnip so early in the season. When bagging them for the Fairs be sure to remove the tops and when they are sliced for the sheep they should be perfectly clean because some bad results have been recorded from the feeding of sandy and dirty turnips. With these in addition to cabbage the exhibition sheep will thrive exceptionally well while away from home.

Before the Show

Having brought the sheep of the respective classes as near perfection as possible, as regards growth and condition, much will still depend upon the way in which they are placed before the judges. Every care should be taken in matching individuals for the flock prize. A prize is often thrown away by putting together animals of different sizes and types.

Again, it would be bad policy to send the best animals away from home the week or so preceding the most important fairs, where the breed comes out in great numbers, and the prizes are more valued. Such a step would be placing the sheep at a disadvantage as they would undoubtedly lose much of their bloom.

As To Selection

If the exhibitor feels convinced he cannot show to win, he should not select the biggest to represent him, but rather those which are the truest to character and type, and likely to attract the attention of breeders, with perhaps more remunerative results than the mere money value of a prize. Undoubtedly many men have injured their reputation by exhibiting animals not true to character when perhaps their flock in its entirety was a very good one.

The True Show-Ring Spirit

The exhibitor has much to consider other than simply bringing forward his sheep in the best possible manner. It must be kept in mind what an exhibition is really for, but at all shows we find two classes of men, one including those who consider exhibiting as a sort of war in which they must strive to beat their competitor regardless of how low they have to stoop to do it, and the other class includes those who are imbibed with the true show-ring spirit and are there to help raise the standard and advance sheep interests in general. The actions of the former class have to be endured to a greater or less extent by those exhibitors who are truly gentlemen. They will mix up their flocks and say disagreeable things, but new exhibitors must cultivate their own minds so these sneaking tricks and unkind words from other hands and lips will not excite or vex them. If you happen to have such a competitor, just consider that you are unfortunate to that extent and by cool judgment and clear thinking try to overcome the public effect of his evil ways. Some men, and old ones too, in different cases, seem to think that their reputation can be built up by untruthful low sayings about other people but such is not the case, and when you hear anyone speaking ill of their competitor just firmly impress upon your mind that the person who uttered those words is not a first-class man whose principles are high and correct. The true exhibitor who is a benefit to the industry and is satisfactory to deal with is the one that strives hard to bring forth sheep of the highest possible standard and by placing them before the public and in the show-ring he helps to raise the ideals of spectators and for himself has the satisfaction of knowing what improvement he has made in his own flock and learns how he can make still greater improvement. Do not let your mind dwell upon what the other fellow is doing, but always do your level best to make your flock and yourself better from year to year. To be what we are, and to become what we are capable of becoming, is the only end of life. Too many exhibitors center their thoughts and plans upon winning over some certain competitor, but that is the very lowest class of show-ring spirit, and when continued it not only upsets those breeders but it does harm to sheep-raising in the section where they are. The exhibitor must say to himself, "I will do the best I possibly can and take care of my own business." The world is always ready to welcome such men and they invariably gain the best standing and prove to be the foundation for all real improvement. Those who splutter out and say they will beat a certain man no matter what it costs are just some specie of human hyena and of course if they have plenty of money they can remain in the business and keep right on doing harm and upsetting young men's ideas. We suppose the Creator placed a few of them among us so when we viewed things correctly we could fully appreciate what is right, just, and uplifting. The unassuming, steady, push-forward young exhibitor can gain a foot-hold and an enviable standing anywhere. Truly the show-ring brings more or less excitement and it is good that we are filled with the spirit to win, but it also is high time to bring into full play the good old Golden Rule, "do unto others as you would have them do unto you," and give it front position at all times in your principles which guide your words and actions toward others. Your desire to win should be based upon your purpose to make your flock as near perfection as possible, and then when your sheep are placed at the top of the class you have great and pleasing satisfaction. Would it really satisfy you to win as a result of some mean act, or to make a sale by saying degrading things about another firm? Would that which seemed gain at the moment be a lasting gain, and would it make your life's work better and more satisfactory to yourself? When you start with a high-class pure-bred flock you have begun the foundation for a life-work and your desire is to make a great success. If you were to start a stone foundation for a large building and wanted it when completed to be a standing monument for your purpose and work, would you some day in a hurry put in some mud or other material which would soon crumble or rot away? The fairs and your work there are only but a stone in your foundation of a life's work in breeding, and it certainly will be best in the end to have left no weak places. Judge things with a consideration of the future and if you do right at all times when it comes sharply to you to do one thing or the other it makes you stronger in the estimation of both yourself and the public. Endeavor to be that kind of a showman that the good men enjoy showing against, and live such a strong and straight show-yard life that even your competitors will admire and respect you and your principles. There is more to consider than just yourself and to-day. Young farmers and breeders who pass the show-yard and pens and notice the high quality of your sheep and your personal manner may be buyers in future years. At the start it must not be the only desire to win, in fact that is what leads many astray. Decide that you will do business squarely and produce the best sheep you possibly can and exhibit them fairly. Sound business principles with a respect for others is the only solid foundation to be laid for the future. That does not mean that energy should be slackened, because success is the result of labor. Work as hard as you can but do everything honestly. The exhibitor who has any other principle may for a while seem to be succeeding but as time goes on, the world will notice that so-called success turning to failure. A good many things will stand in fine appearance when all is well but when real tests come it is the sound, high-principled exhibitor that stands and the others that fall. Work hard against those that really do the industry harm by inflating jealousy, etc., but always assist the man whose aim is for the right goal. The best breeders must unite to advance general interests. Selfishness is not included in the true show-ring spirit, and even with strong men it is a gain for today and a loss for tomorrow. The true exhibitor goes to the show to build up his reputation and help advance his breed and the industry. When we have studied deep down into the foundation principles which when combined will bring about these results we will have learned what if applied will cause us to have the true show-ring spirit. It is important and desirable that agricultural exhibitions be held to help general breeders to have higher ideals and also to stimulate the exhibitors to steady improvement in their flocks. Therefore it is the personal duty of each exhibitor to do his full part in assisting to bring about these results. Will any other than straight principles make object lessons which if copied after will make real advancement? Truly the exhibitor has first in mind his own welfare and the spreading of his reputation, but a broad mind, a generous heart, straight dealing, and good sheep are all that will ever bring him lasting personal gain. Do well by others and they will do well by you. The true show-ring spirit is broad and really helps everyone in the sheep business and gives spectators and beginning breeders high ideals. Talking with some exhibitors is an inspiration and they fill your mind with great and good thoughts, while other men will in the end make you feel mean and selfish; the former have the true show-ring spirit and the others have not. Try to be of the former class.

"CLOVER HILL'S 4047 A. S. A. 261015.

First Prize two-year old Shropshire ewe in Iowa and American-bred classes, Iowa State Fair 1909, bred and owned by exhibitors, Chandler Bros., "Clover Hill Farm," Chariton, Iowa. Also one of the Champion Iowa-bred Shropshire Flock, 1909.

Management of the Breeding Flock

The breeder's year really begins when the ewes are mated in the autumn, so we will first refer to the mating season. During recent years a great deal of attention has been given to **flushing breeding ewes.**

This is a practice which is now being taken up by nearly all leading breeders in America, and for many years has been practiced in Great Britain and other foreign countries. The term "flushing" is applied to having the ewes rapidly gaining in flesh at the time ram is turned with them for mating.

There is abundant evidence that "flushing" hastens forward the mating time. It has been fully demonstrated that "heat" in animals is brought about through the action of an internal secretion elaborated by the ovaries (or organs which give rise to the ova or female germ cells). It would appear, therefore, that the artificial feeding which would be given the ewes at this time exercises a stimulating influence on the secretory action of the ovaries, while at the same time causing the graafian follicles (or ovarian vesicles which contain the ova) to reach maturity more rapidly, and a larger number to discharge during the early "heat" periods of the mating season. Besides causing the ewes to take service of ram at an earlier date, this additional and fresh feeding tends to increase the number of lambs dropped. Perhaps the results to be obtained from "flushing" have never been fully explained to all, therefore we wish to give the details of some experiments made by Francis H. A. Marshall, lecturer on the "Physiology of Reproduction" in the University of Edinburgh, Scotland.

In 1905 there were three pens of ewes in the experiment. One pen was fed on only grass during the summer months previous to mating. During the three weeks they were with the ram they received a full supply of turnips, and during pregnancy received dried grains and turnips, and were fed on "lamb food" about three weeks previous to lambing. The rams were fed on bruised oats during the time they were with the ewes. From this pen of ewes 12½ per cent had triplets and one ewe had four lambs. In this instance the exact percentage of lambs was 191.5. None of the ewes were barren and none aborted. Unfortunately, however, not all of the lambs could be reared, so the number still living at about one month after lambing was reduced to 183 per cent.

The second lot of ewes were given Bombay cake (a mixed feedstuff), bruised barley, a small amount of linseed cake, as well as turnips during mating time. Previous to this they were

were fed only on grass. Some turnips were allowed during the period of pregnancy. The rams received the same feed as the ewes. Thirteen and one-half per cent of the ewes produced triplets. None aborted or were barren. The ewes were all 3 years old and produced 193.75 percentage of lambs.

In the case of the third lot of ewes, they were placed on a fresh better pasture just previous to mating time, and from this time on until the middle of April they had good pasture, a reasonable allowance of turnips and all the cut hay they would eat. The rams received no special treatment. Out of 184 ewes, 23 had triplets, 2 ewes were barren, and 1 aborted. Altogether this pen produced lamb percentage of 196.

Further Mr. Marshall states that it is obvious that lambing returns as a whole confirm the conclusion that extra feeding at mating time results in a larger crop of lambs at the subsequent lambing.

One correspondent stated that he put 60 ewes on rape and clover at mating time and that 90 per cent of these had either twins or triplets at the ensuing lambing, far exceeding the other sheep, which were treated differently.

There is distinct evidence, also, that the barrenness percentage is less with ewes which have been specially fed in the way indicated.

These facts should be of great interest to every sheep breeder, because it is quite an item to produce a large percentage of early lambs, and the outlay of additional feed is very small.

The result of this practice has, of course, been perfectly well known to numerous individual flockmasters, who have consistently "flushed" the ewes on their own farms for a period of many years. It is surprising, however, that although feeding experiments upon mutton and wool production have been described in various agricultural publications, no systematic investigation dealing with the effects of different methods has ever been recorded. The complete absence, so far as the writer is aware, of any definite records on this subject may perhaps explain the want of knowledge of the results of extra feeding among large numbers of sheep breeders who have never adopted this method of increasing the fertility in their stock. In England the subject seems to have received even less attention than in Scotland, and in our own experience the practice of "flushing" is often deprecated by those who have never tried it. So far as we have been able to ascertain, those who have adopted the practice of flushing their ewes are satisfied that the extra cost which such additional feeding involves is more than repaid by the larger crop of lambs which is produced.

All who have flushed their ewes have found it very profitable, and, although not all the feeds used in the Scotch experiments are procurable in America, it is unnecessary because other feeds will answer the same purpose. By referring to the third lot of experiment ewes it will be seen that it could be carried out very simply on any American farm. Nearly every farmer has a fresh patch of second growth clover, and clover can hardly be excelled. Rape is good in addition, and if it were in corn or alone joining clover, you have an ideal combination. To turn your breeding ewes on a patch of clover and rape would immediately affect the ewes in the manner mentioned in the first part of this article. The ewes would come in "heat" very soon, and mating under such conditions would bring about the desired result of a large number of lambs.

The best results ever obtained at "Clover Hill," in Iowa, were from ewes turned on such a patch of fresh clover. We erected a few temporary "V" shaped troughs and fed them a liberal allowance of chopped pumpkins sprinkled over with salt and oats. The rams received the same treatment and were allowed to run with the ewes. As soon as the ewes had all taken the ram and refused the second time, the ram was taken from the flock. There were two sets of triplets from this lot, and nearly all raised twins; not a single ewe being barren. Lambs were dropped very closely together as regards date, and this assists the breeder in having a uniform bunch of lambs, which is very desirable. This is one step toward the production of "more better sheep" which should not be overlooked by anyone. Every farmer can "flush" his ewes in this way and once you have practiced it you will always continue. Attending to such details is what brings success to some breeders while the inattentive are wondering but never succeed as they would like to.

Later Management

After the ewes have remained to the service of the ram they should be turned on a good old turf. Blue grass pastures fairly well grown are the best after the heavy frosts have come on. As winter approaches a little grain could be given and when snow has come of course hay feeding must begin. Clover hay is by far the best hay produced in the Middle West. Alfalfa is fully as good but is not raised very much

East of the Missouri River. Breeding ewes should be kept in good thriving condition but not necessarily fat. They must have plenty of exercise so give them as much range as possible. When winter has set in of course the flock must be protected but do not shut them up in some small dark ill-ventilated and poorly-bedded shed. A sheep shed should be large, the roof high, and in some climates it is best to have the south side open. Unless you live in the north where winters are severe the flock should be turned out every day so they may have exercise. Scatter a few little bunches of corn fodder or hay on the side hill out of the wind and such practice will do the flock no end of good. Keep the shed clean and well bedded with oat or wheat straw. Remember that the ewes have their own bodies to keep up as well as the unborn lambs and they must have bone and muscle producing food. Timothy hay and corn produce fat so they should not be used. Clover hay, and pea straw are good rough material, while oats, bran, oilmeal, etc., are the proper trough feeds. Oats and bran in equal parts with about one-fifth the amount of crushed oil-cake is a most desirable mixture but as lambing time draws near the amount of oilmeal could be slightly increased. Properly fed ewes go through the lambing period in good form and drop strong lambs. Just before lambing each ewe should be placed in a small pen by herself and allowed to remain there until the lambs are two or three days old.

Merits of Sheep Forage and Feeds

Clover Hay

Clover contains the greatest amount of feeding value when cut just as the heads are in full bloom. When in good condition it furnishes a large amount of protein and ash essential to thrift in the breeding flock. It helps to fully develop lambs which are to be retained as breeders because it builds strong bone and a large framework; it gives proper nourishment to the ewes carrying unborn lambs and gives more favorable returns than any other from of roughage to breeding sheep of all classes. Clover which has been poorly cared for and weatherbeaten becomes very harsh and, in addition to losing much of its feeding value, it is liable to cause stomach trouble. Too much attention cannot be given to properly gathering it so all the little leaves will be in fine condition.

Corn Fodder

To obtain the largest amount of nutriment in the crop, the seed should be planted very thickly so the ears will not attain full size. The best time to cut for fodder is when the kernels are just past the glazing stage. If cut earlier the plant and ear would contain more water and less feeding value. Much attention should be given at cutting time in order that the stalks be placed erect and in very large shocks. This diminishes the loss of leaves by the wind and the detriment done by rains. Then it will be a valuable winter roughage for the flock, although the dry matter in fodder does not give quite such exceptional results as when the plant has been cut for silage, those breeders who do not have silos must not overlook the importance of having some good fodder.

However, it should not be fed exclusively, but can be rated very highly when a small allowance is fed daily. The leaves are a coarse hay of high feeding value, and the ear, having been left in the husks, is eaten with greater relish than the hard grain which has fully matured and been placed in a bin. There is no much better forage for sheep, but care should be taken that they do not get too much grain. Fodder is best fed outside where it tends to give the flock their needed daily exercise.

Timothy Hay

Timothy hay with its stiff, woody stems, yields a very small amount of forage and should not occupy an important place on the sheep farm.

Alfalfa Hay

In making alfalfa hay the greatest care should be exercised in saving the leaves and finer parts, so easily wasted. The possible loss from careless making is great, but when properly cured, alfalfa is very palatable to sheep. Perhaps it will give slightly better results to fattening sheep than clover. Where it can be successfully grown it finds great favor for this purpose, but in the corn belt the amount of special work required in its production is much greater than with clover. The tap root of alfalfa reaches many feet into the soil, thus indicating that the plant must have a subsoil through which roots may pass and water should not be near the surface. For breeding sheep the feeding value is not much different than clover.

Millet Hay

Millet hay is not considered a good rough feed for sheep. If it is to be used for hay it should be cut when just coming into bloom to avoid the formation of the hard seeds which are nearly indigestible by live stock. In too many instances it is cut late, and then when the hay is used entirely it is apt to be very injurious. The principal objections are that it causes increased action of the kidneys, also scour. More care is necessary in feeding millet than any other coarse fodder. If it is to be fed, it should be in limited quantities and not continuously. If you have it on hand and do not care to purchase clover or other hay for whole allowance it would be advisable to at least feed an amount of clover equal to that of the millet.

Oat and Pea Hay

The prominent characteristic of the field pea is its large content of protein or bone and muscle building material, this richness of protein rendering it particularly useful for breeding ewes and growing lambs. When grown with oats it is a feed that merits the consideration of all flockmasters. We have received good results from feeding oat-and-pea hay, the seed having been sown at the rate of about two bushels of oats and one bushel of peas per acre. Seeding can be done practically as early as oats alone. Just as the oats are turning yellow at ripening time the pea seed will also have passed the milky stage and it is the right time to cut it for hay. This combination forms a forage of high nutritive value much appreciated by sheep. Attention should be given the crop at this time because if the crop were to be cut too green the seeds will mold and lower the feeding value. Properly cured oat and pea hay is a most excellent feed during winter months for the breeding flock, and it makes them thrifty rather than fat. The yield per acre is quite heavy and the mixture affords the proper winter change from the regular feeds. This same mixture can be grown to feed exhibition sheep and cut when desired before ripe for feeding inside.

Corn Silage

With American breeders the use of some kind of succulent feed nearly the whole year round has become quite general. It helps to keep up the appetite and general condition of our flocks. Although roots are not so successfully grown as in Europe, corn takes their place and furnishes a larger and cheaper supply of food material from a given area than any other crop. It will yield about twice as much dry matter as a crop of roots grown on the same land, and it has been found by feeding experiments that the dry matter in corn silage gives as good results as that in roots. All breeders who can should have a silo because silage is so much more palatable to sheep than dry fodder and they will consume a larger amount of dry matter in that form and it is more easily digested. The use of silage as a succulent food for sheep has given most favorable results and experiments in fattening sheep have shown that corn silage gives better results than rutabagas or Swede turnips. We believe that it is the most desirable succulent food produced in the corn belt for both breeding and fattening sheep. Most of the adverse reports on silage are due to the use of green immature plants and such silage apart from being sour is of very low feeding value as compared with that made from the crop which was well matured before being harvested for the silo.

Indian corn is best suited for the purpose because when cut it packs very closely in a solid mass and keeps well. Like roots silage makes a watery carcass which is soft to the touch and this is a desirable condition in all breeding sheep also fattening ones during the early stages of that process. For breeding sheep the less tense flesh, a natural result of silage feeding, is more conducive to vigorous young at birth and to their hearty maintenance afterward than dry feed continually throughout the winter. Feeding only dry forage tends to produce a dry firm flesh—a condition certainly not conducive to the highest degree of health in the flock. Too many flocks give this dry harsh appearance in winter and it proves a loss to the owner in both the lamb crop and the wool. Silage tends to keep up the same condition that is noticed when the sheep are out on green grass in the summer, and the cost of its making is not very great. It also takes the place of much grain which would otherwise be required.

Indian Corn

Corn as a grain is much relished by sheep and is more palatable than others which turn to a sticky mass during mastication. It has no equal for fattening but owing to its low per cent of protein and ash, it is not well suited for developing young or breeding sheep which require food that will produce bone and muscle. As it is raised by nearly every American farmer it is a common feed, but we believe that a great number of failures to obtain the greatest results from breeding flocks are due to the liberal use of corn. Ewes thus fed will fatten instead of properly nourish their lambs and then the owner wonders why they are not as healthy and thrifty as they should be. Liberal feeding and proper feeding are often quite different because no properly fed breeding flocks receive a liberal allowance of corn, unless it is in conjunction with oats, bran or some such feed.

Oats

Oats are perhaps the most desirable of all grains for the breeding flock because they produce growth rather than fattening. Oats contain a much higher proportion of bone and muscle-producing nutriment than corn and in itself is quite a well balanced ration. Especially is this true when the oat kernel has a small hull. If the oat crop is of poor quality and the hulls are woody it is well to add some bran or oil cake to the ration. In some cases oats have given better results when ground but we do not believe this is necessary or profitable for sheep feeding.

Wheat

Wheat for sheep feed is much more balanced than corn, and contains a larger amount of that nutriment which is required for the full development of lambs and also contains nearly as large a per cent of fattening material. In most cases the price of wheat forbids its use for live stock feeding but those sheepmen in sections where wheat is plentiful are fortunate. It gives best results with sheep when fed whole but as it is a strong feed it should always be mixed with other grains. Some sheep breeders have obtained excellent results from feeding wheat alone but such use can not be recommended generally.

Bran

Wheat bran carries a large amount of crude fiber but it is very desirable for a mixture with grains. It produces the effect of a mild laxative which is quite beneficial. It contains a large amount of muscle and bone-building material and gives most excellent results when fed to growing lambs, breeding ewes and rams in service. For ewes with young lambs it is a leading feed with the best breeders because it not only furnishes bulk but large quantities of protein and ash which are so much needed in the formation of milk. Farmers who have plenty of corn but no oats or wheat can make a good ration by mixing bran with it. Sheep which are housed being fed for exhibition purposes should always have bran in their ration especially on account of its bulkiness and laxative effect upon the digestive organs. Breeding sheep should always receive some bran unless they are out on grass.

Oil Meal

Oil meal is the residue after ground flaxseed has been subjected to great pressure for the purpose of removing the oil. At first it is in slabs about an inch thick, perhaps fourteen inches wide and about two feet long. For feeding, these slabs are reduced to the size of hazel nuts and this is called "nut cake" and is the most desirable size for sheep feeding. Usually it is ground to a meal but that forms a paste in the sheep's mouth which is undesirable. Oil cake or meal is a very healthful feed and places sheep in a fine general condition with a pink skin, oily fleece and good quality of flesh. It has a most beneficial effect upon the digestive organs and the flock always profits by having a small portion mixed with its regular winter grain allowance.

Practical Sheep Barns

Proper shelter for the flocks adds greatly to their thrift, while improper sheds are in many instances not much better than nothing. To economically build barns which are correct at all times and for all purposes should be the desire of flockmasters. In the summer the pure-bred breeder wants a barn properly arranged for the well doing of sheep which he will be preparing for exhibition purposes, when cold disagreeable weather comes in the autumn all breeders want a place for the flocks to lie inside and have a little clover, during the winter ample shelter must be provided from heavy storms and a dry clean place is needed for feeding, in early spring a warm dry place is needed for the ewes and young lambs. Those are the needs for shelter. Then comes the matter of feeding and not only must proper feed troughs be provided but the rough feed should be close at hand. Arrangements should be made for some means of sorting the sheep without turning out of the barn. Some way of easily loading into and unloading from a wagon should be provided. There should be some space for mixing grain, to place buckets and all other loose necessaries about such a barn. The shepherd should have a room of his own, so he will be near the sheep at all times, and especially during the lambing period.

We have been many years in planning a barn to meet all such requirements, and we are now fully satisfied with those we have erected at "Clover Hill Farm." The one photo is of the outside of one of our barns, and the other photo gives an idea of the trough and pen arrangement inside. We do not believe in the low poorly-ventilated sheds found on so many sheep farms, and furthermore we want all hay and straw above the sheep where it is easily obtainable regardless of weather. Rough feed and straw, being kept from the weather in this way is also very bright and there is no loss such as is experienced by stacking outside. Such a barn lessens the shepherd's work and saves much general expense to the

OUTSIDE VIEW OF A PRACTICAL SHEEP BARN AT "CLOVER HILL FARM"

owner. Large doors next to the roof-peak open down to permit hay etc., to be sent in by means of fork or sling on a steel track. Such entrance is at both ends, the carrier running from either end to center of hay mow. A large feed way, about 6x4 feet, extends from top of mow to shed below and this being arranged in center of barn lessens the winter work when feeding. The floor of the mow is 7-inch matched house flooring so no dust can possibly fall on the sheep below. This barn, 80 feet long and 36 feet wide, has hay room sufficiently large to hold all rough feed needed during a year for the 300 sheep which the shed below will hold. In the inside photo you will notice ladder in middle of barn going up this feed-way down through which the hay etc., comes. The hay drops into the alley-way 4 feet wide which you will see runs full length of barn. In addition to being a place separate from the sheep through which all feed can be carried the alley-way is very convenient and clean for visitors to see all the sheep. Feed racks are made according to our own desires and ideas, being a combined one for both grain and hay. The bottom is a board ten inches wide, having a 3-inch piece nailed on each side. This bottom will then hold grain roots and the like. The upright pieces nailed to this are 1x4 inches and 18 inches long nailed to a 1x4 top piece running lengthwise. That makes a hay-rack 10 inches wide and 18 inches high which is sufficiently large to hold all hay the flock will eat during a night. From experience we have learned that it is poor policy to place large quantities of hay before sheep. Feed just what will be regularly eaten and then the rack can be cleaned for grain.

If for field ewes or rams and they are not fed grain, the troughs even then should be cleaned each time before putting in fresh hay.

All wood material used in these racks, and for all other inside work should be planed smooth so the fleece will not be continually roughed up. We have found these to be very desirable and serviceable racks and they have been copied by many "Clover Hill" visitors. We erect them stationary full length of barn on each side of alley-way (or walk-way), except where little doors open into each pen. This barn being 36 feet wide gives us a 16-foot space on each side of the 4-foot alley-way. We make these trough-racks of the right length to extend from the outside wall of barn to the trough next to alley-way but do not fasten them so securely but that they can be removed so pens can be made any desired size. Regularly we place them crosswise 16 feet apart and that makes pens practically 16 feet square. The upright pieces in troughs are 10 inches apart so a sheep can get its head in and out easily. We believe that is the correct space between

INSIDE VIEW OF A PRACTICAL SHEEP BARN AT "CLOVER HILL FARM."

uprights. As there is trough space on three sides of each pen it leaves space for about 40 sheep to eat in each pen 16 feet square which is as large a number as would be put there for shelter. There is a little gate on hinges from each pen to the alley-way and that makes easy what sorting or changing you might wish to do at any time.

In the outside photo at the end of barn you will notice the open door about three feet from the ground. It is at the alley-way end and when a wagon is backed up to it the wagon-box bottom is just level with barn floor thereby enabling us to drive sheep from any pen through the alley-way into the wagon without lifting. The barn foundation is of crushed rock and cement, 5 feet high, and the earth was not fully graded up to this end of barn. To each 16-foot pen there is a large door from the outside, also a window. This permits the driving of sheep in or out of any pen without disturbing the others and is also convenient when hauling the manure from each pen.

During the heat of summer a blind is placed over the windows to keep the sun's rays out and both top and bottom of door are fastened open and a slat door is placed on the inside. The air must be let right down to the sheep, it being far from proper to follow the general practice of opening the top doors and keeping the bottom ones closed. The floor of hay mow is 8 feet from barn floor, so there is great room for fresh air. Sheep need clean dry bedding and plenty of ventilation. With all these doors and windows and a high ceiling, it is a model barn for summer. For winter the top doors and windows admit ample sunshine. On this side of barn at further end upstairs you will see a window and chimney. They are from the shepherd's room which is plastered painted and fixed up like a house. The stairway from it comes to the barn alley-way. The space, 16 feet square, under his room, instead of being a sheep pen, is his for feed, etc.

The outside wall of barn is of the best quality of matched drop-siding without knots making it practically air-tight. The barn when full of sheep can never be completely closed without soon becoming too warm. We believe in a tight barn and then use windows for ventilators. Cracks between the boards are improper ventilators although the way some barns are built would indicate that their owners thought cracks were just right. When a barn is built with best drop-siding and has hay above it can be easily kept warm enough for young lambs in nearly any weather.

We have used these kind of barns at "Clover Hill" for several seasons and have had no reason to even slightly change them in any way.

The Shepherd's Work

The shepherd's work is an endless task and whether he is owner or hired man his thoughts must be centered on the welfare of the flock. If you are the owner and are working with the sheep there is no doubt but that you are deeply interested in the success of the flock but when you hire a shepherd be sure that you get a real one that knows a great deal about sheep. Every day in the year there is something to do and it must be done well.

When the ewe flock is sorted in the autumn to be bred to certain rams because you believe they will produce the highest class lambs from such mating you do not want a man who is so careless that the ewes might be bred to any ram just so they get with lamb. It is the good shepherd who plans out what should be done at all times and then combines perseverance with thoughts and brings about everything as planned. Numberless flockmasters and shepherds know what should be done but that number would be greatly reduced if it were to be confined to those who put a large portion of their good ideas into practice. There is just as much difference in the management of flocks as there is in flocks themselves. Proper caring for a good foundation flock leads on to success but it doesn't matter how good the foundation may be if they are poorly cared for—failure to some extent will be the result.

The good shepherd is continually thinking about how he can improve the flock and surrounding conditions. In this way he is also improving himself and his income. Breeding and feeding are his two principal studies and they must always be combined. In a pure-bred flock he must be conversant on all pedigrees of noted rams and ewes and upon the merits of the leading flocks. This knowledge will enable him to make selections of sires or ewes for flock addition that will probably produce lambs of the same high type. This enables you to properly found a flock and keep it making rapid progress upward in quality.

Sheep properly fed will be fully developed, and the good individual well developed is par excellence. To assist nature must be your desire and not to work against her. When a sheep or any other animal does not receive more than enough food each day to simply replace the strength which is expended in living and exercise it becomes "stunted" or is too small and lacks bloom.

Good feeding does not necessarily mean a heavy allowance of fat-producing food such as corn, etc., but in breeding sheep it implies plenty of food which will build up muscle and bone and tend to growth, not fat.

Oats, bran, linseed cake, clover hay, etc., have helped bring our improved breeds of sheep to their present high standard of perfection and will assist us in raising this standard still higher.

Careful methods of breeding give us the proper material to work on and good feeding enables us to assist nature in developing it fully. If your sheep lack in breeding, the same amount of feed will not give you as large returns as it otherwise would, and on the other hand it doesn't matter how well bred a ram or ewe may be if it isn't properly fed it will never be as strictly first-class as it might have been. Therefore we say that the shepherd must have a thorough knowledge of the best blood of the breed he is caring for and know how to feed for full development.

The flock that is well and properly fed has very few ailments so when you learn to feed well you have torn out a number of pages in your "doctor book." The sheep's system which is properly nourished by clean food will rarely get out of order and that covers it all. Coarse, half-spoiled old hay and ice-cold water in the winter time would nearly give anything indigestion, and possibly chronic, too, but that would not be good feeding. A barren old pasture or one with stale grass and dirty water might not make nice fat lambs in the summer, nor you couldn't expect it to. Bred ewes in the winter placed in a roomy well-ventilated shed, bedded with clean, dry straw, and with a small allowance of oats and bran, and well-cured clover and corn-fodder for rough feed will thrive well and produce heavy fleeces and a good crop of strong lambs. If lambs in the early summer are turned on a patch of rape, kale or some good fresh feed they will thrive and be a credit to any farm.

There are always two ways to do a thing, and you want to be sure that your sheep are bred, fed and managed in the right way.

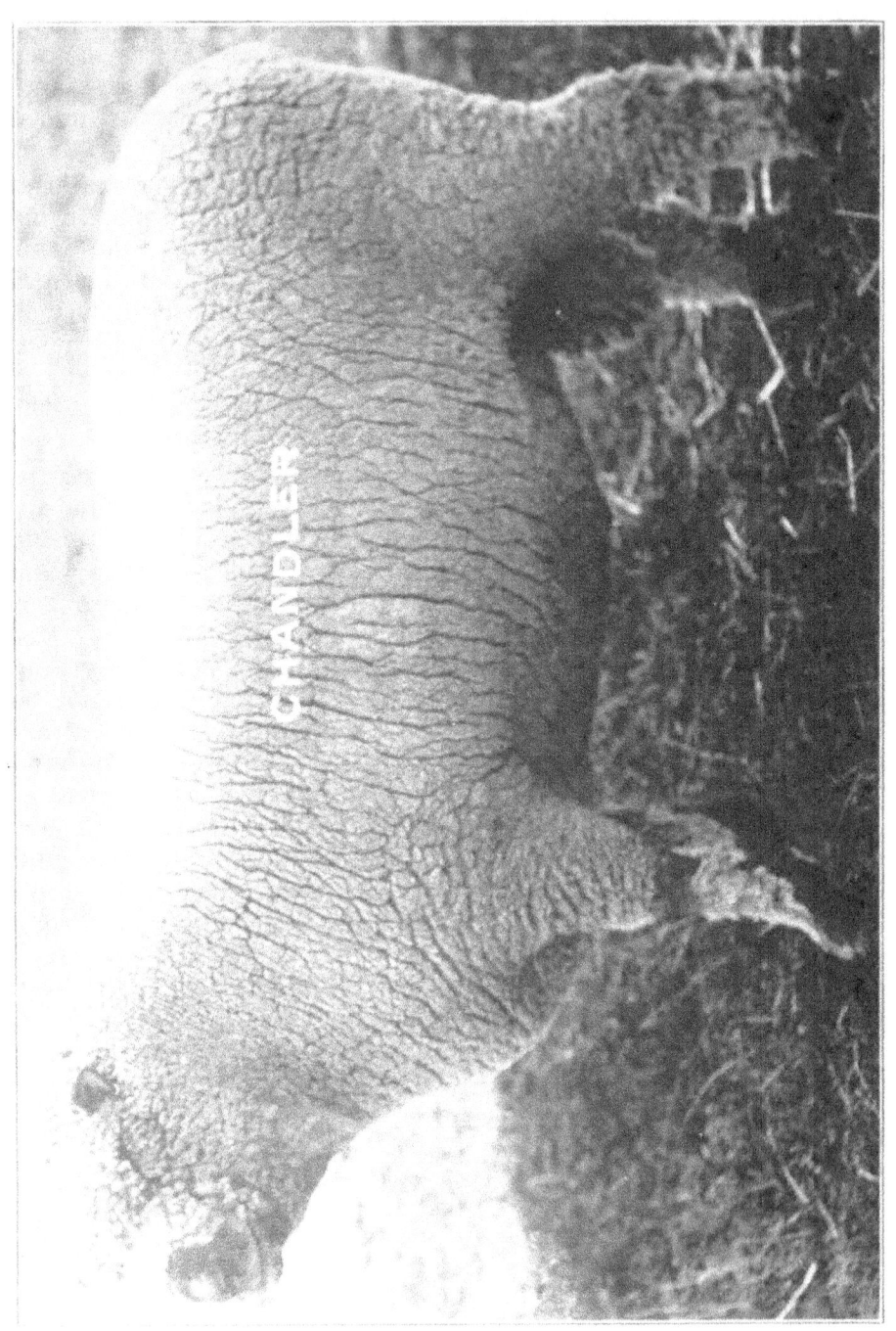

"CLOVER HILL'S WILLIAM THE CONQUEROR" (29075)
A sire now in use in the "Clover Hill" Flock of Shropshires.

Necessary Precaution in the Lambing Fold

Should any ewe die abort or strain after lambing she should at once be removed to a safe distance and the wood-work and pen it occupied must be thoroughly disinfected with carbolic acid or other disinfectant, and all the litter, etc., burned. It is also wise to have in the lambing fold a tub of live lime with an empty tub and shovel alongside. All cleansings etc., should at once be placed in the tub and a shovelful of fresh lime thrown over it. By this simple process the lambing fold is quite free from unpleasant smells and the possibility of contagion from unhealthy matter is greatly minimized. These may appear, to some, unnecessary measures to adopt but those who have noted the disastrous results brought about by carelessness will readily see the desirability of carrying out these suggestions.

Much of the so-called bad or good luck is usually traceable to the treatment the ewes have received during the pregnant period and a high rate of mortality amongst ewes and lambs is often the result of their being kept on cold wet or barren soil.

Assistance in Lambing

As a general rule it is better not to interfere too soon but it is easy to tell when to assist Nature. In all cases where help is given, carbolized oil should be freely used on your hands and arms before operating, and in bad cases and when a ewe has been assisted with decomposing lambs the carbolic oil should be poured into the vagina by raising the hind legs of the ewe and allowing it to flow in. Great care should be exercised in such cases and when disease is rampant to wash your hands, etc., with carbolic soap after each operation and freely use disinfectants, in fact, you should never go from a bad case to assist a healthy ewe without taking every possible precaution to prevent contagion. All the surroundings such as litter and food of any case where disease was apparent should be removed and burned and the pen thoroughly disinfected without delay. A barrel with a hinged lid with some nice dry hay in it is a capital place to put a sick or weakly lamb in and is really much more effective than placing it near a fire.

For a very weak lamb, a large flat India rubber bag filled with hot water is far preferable to fire warmth and has been very successful in saving life.

EWE LAMBS "CLOVER HILL STYLE" A. S. A. 23767, "CLOVER HILL PRIZE" A. S. A. 23768.

"Clover Hill Style" was practically female Champion of 1906 and at finish of show season won First Prize at International Show, Chicago. Exhibited by Chandler Bros., "Clover Hill Farm," Chariton, Iowa.

Treatment and Feeding of Ewes and Lambs

It is important to get the ewes with lambs out of the shed pens as soon as possible except in cases where lambs may be too weak. Sunshine and fresh air are as essential to their growth as food is. Of course the ewes must be fed so they will give large quantities of milk. If you will notice the condition of young lambs in flocks where different methods of feeding are practiced you can readily see what proper feeding to the ewe does for the lambs. There is marked difference when one lot has been fed corn or corn fodder with timothy hay and the other fed linseed-cake oats and bran, the lambs from the former being weak simply because they have never received sufficient nourishment from the dams. The ewes were unable to give it to the lambs because their own food did not give such nourishment to their bodies. Corn and timothy hay do not build bone and flesh, they are fat-producers. Linseed-cake oats and bran together with clover hay stimulate the milk flow and indirectly produce large vigorous lambs. The gain from properly feeding the ewes at this time is so great that all breeders should give the subject their careful consideration. Where rye can be raised sheep breeders are fortunate because a patch of green rye for the ewes and lambs is valuable in addition to the dry feed. That food combination could hardly be improved upon. The "youngsters" like a bite of green stuff and soon will be eating quite a little. If there is no rye it is well to have a fresh pasture that comes up quickly in the spring. This green food assists both ewes and lambs in thriving better than if they were confined to even the very best of dry feed so it is good policy to get them all outside as soon as they are strong enough and the weather will permit. The lambs can also be pushed on by giving them a small amount of grain by themselves each morning and evening. For this, a "creep" can be made easily and cheap. If the flock is being shedded each night a corner of the shed can be used for it. A simple creep is made by taking two 1x6 inch boards and placing them far enough apart to make a partition which ewes will not jump over, then nail slats on far enough apart so the lambs can go through but the ewes cannot. If the flock is placed only in a pen at night a corner of the lot could be taken the same as in the shed. Put a little trough in there and the lambs soon know what it is meant for. Feed to make them grow, not fatten, and when the lambs are yet real young it is well if the oats in the ration are crushed. If lambs get this grain a liberal amount of milk from their dams and some grass or rye you may rest assured that you will get the most from your flock. Lambs which have been well fed in every way always mature to be much larger stronger, and with heavier fleeces than the others.

Ear-Marking the Young Lambs

Some pure-bred flockmasters experience difficulty in keeping the young lambs properly marked so it can be readily told which dams they belong to. Of course some people can tell just which lambs belong to different ewes but to depend upon that very long after the lambs are born is very uncertain. However, the cartilage of a lamb's ear until it is quite well grown is not very strong and tags should not be put in the ears because sore or improperly formed ears might be the result. Therefore, the use of tags for real young lambs is not to be recommended nor should anyone depend upon remembering where every lamb belongs but the safe and sure method is to punch small notches in the youngster's ears and then the tags can be correctly inserted whenever you are ready.

The ordinary ear punch is used for this purpose and we herewith give an illustration of one system of marks and their meaning. If the numbers on your ear tags do not run up very high, the same number could be notched in the ear as the tag you wished to later insert. If the flock is a large one, a note will have to be kept of numbers in order that no mistake will be made. For instance: Suppose this year's lambs would be tagged from tag 500. Then the lamb notched 1 would require tag 500, 2 would require 501, etc. There would be no conflict in starting each year's lamb crop with notch 1 because in the autumn they would be ear-tagged and there would not be anything but lambs without tags. This system of notching is also very beneficial when older sheep lose their ear tags, because these notches will identify them. Suppose a three-year-old ewe lost her car tag, you could get her notch number and age and then look up the ear tag number of such a notch that year. That would enable you to get correct duplicate tag. We believe this system would profit all pure-bred breeders and the illustration makes its detail quite plain. If you wanted to number a lamb 253, a hole through the ear and a lower notch next to the head would be punched in the right ear and a notch in the tip of the left ear. These marks are lasting and when once you become accustomed to the number indicated by each notch, it is a very simple system. These small marks do not injure the young lamb's ears and it is much better than to insert heavy tags before the ears are fully developed.

Dipping

As soon as practicable—say when the ewes are shorn and before the lambs are weaned they all should be dipped and to entirely keep ticks off the process should be repeated in autumn. The object of dipping is to destroy the parasites in the fleece, to kill off any young insects which may afterwards hatch out and to protect the sheep from subsequent attacks.

Experience has taught us that sheep thrive much better when their skins are clean and it has been clearly proved that a good Dip increases the quantity and improves the quality of the wool. It is absolutely impossible for lambs infested with ticks or other parasites to thrive properly, owing to the constant irritation set up. In trying to get relief lambs often nibble at the fleece and swallow small portions of wool, with fatal results. A good and regular system of dipping the entire flock is money well expended. Hence most leading flockmasters dip twice in the season, once as indicated, and again in the autumn.

The modes of dipping are various. For small flocks the hand-bath is in general use, but the swim-bath is by far the best when flocks are large enough for its adoption, as this system gives much less trouble, saves labor and expense, and the operation is far more effectual.

Amongst the various Dips on the market to-day "Hylo Dip No. 3" is considered supreme, being most carefully prepared by The Hylo Company, Marshalltown, Iowa, who are qualified men of large experience whose sole aim during the last number of years has been to produce a First-Class Dip. "Hylo Dip No. 3" is permitted by the United States Department of Agriculture to be used in Official dipping. It is non-poisonous, and absolutely the most effective that has ever been used at "Clover Hill Farm."

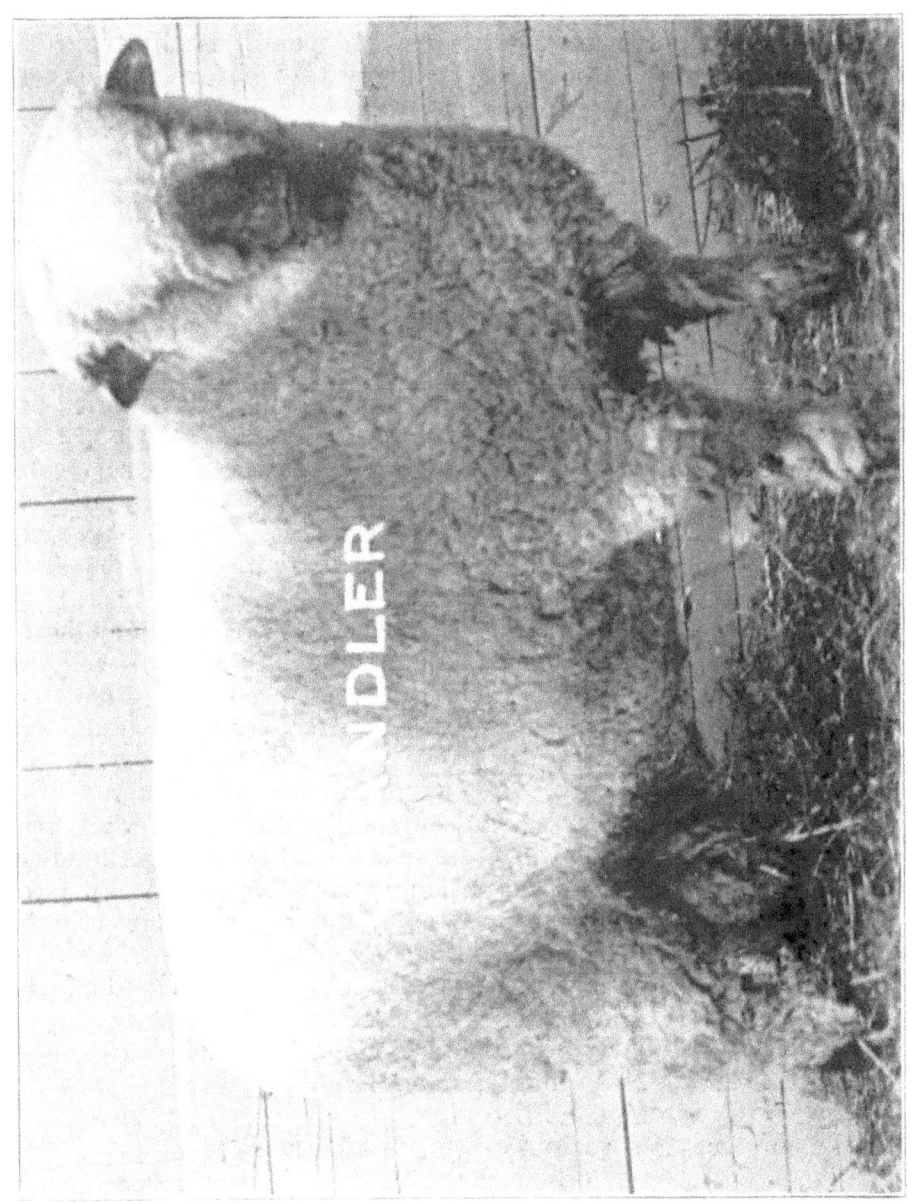

"CLOVER HILL'S FIELD MARSHAL" (29829)

First Prize Shropshire ram lamb, Iowa and Minnesota State Fairs, First Prize ram lamb and Reserve Champion ram any age
International Live-Stock Exposition, Chicago, 1909, owned and exhibited by Chandler Bros., "Clover Hill Farm," Chariton, Iowa.

Weaning

Weaning if not conducted with care and with proper forethought will often lead to a derangement of the system of the lambs.

In some cases, at a very early period, the lambs are separated from their mothers and at once placed on an old pasture, which is perhaps burned up and totally unsuitable to the tender stomachs of the lambs. The sudden transition from the milk of the ewe to the dry summer food is beyond the power of their digestive system. The new food is not properly assimilated, consequently general derangement of the stomach and system immediately follows.

To prevent these evil results, the food supplied at weaning time should be of a highly nutritious quality such as can be easily assimilated and if the weather is hot and dry a plentiful supply of clean water should at all times be available.

The date of weaning depends on the locality and the breeder must be guided by circumstances. Early weaning is in most cases to be recommended for the following reasons: Keep is plentiful at this season, and this affords an opportunity of giving the lambs the best pastures and putting all the ewes into one field instead of being all about the farm robbing the lambs. In some districts weaning is so late that aftermath clovers are available but in the majority of cases this will not be so, and if rape, or other green crop be to hand, so much the better as the object is to minimize the loss of the milk as much as possible. A little grain should be given but it must not be of a heating nature, probably nothing surpasses extra good crushed oats with bran and oilmeal. The lambs should as soon as rape can be got be put on the arable land and pushed forward in a healthy natural way avoiding an undue proportion of artificial food. As the harvest is cleared the aftermath clovers afford a good change for the lambs. Close folding if possible should be avoided as it tends to fatten and not to develop muscle and strength which should be the object in a breeding flock. By this is meant that the hurdles should not follow close upon the lambs but that they should be allowed to roam at large over the field.

The experience of breeders during the last decade seems to point to keeping the lambs from the period of weaning right through the autumn on arable lands eating a variety of green foods--turnips and young clovers and not on old pastures.

"CLOVER HILL ENGLISH QUEEN" A. S. A. 234722.

First Prize yearling Shropshire ewe at English Royal Show 1906, also a leading winner
at American Fairs, exhibited by Chandler Bros., "Clover Hill Farm," Chariton, Iowa.

To carry this out the breeder must exercise a little fore-thought and arrange for a succession of kale, rape and other suitable foods. This can be easily done by planting so much winter rye in the autumn, following up with early cabbage, planted in March or April according to the weather, the drilling of the early Enfield cabbage at intervals during the spring and summer months assisted with white turnips rape and kale in suitable quantities. A large flock can be kept in this manner.

Noting the Ewes Which are Best Breeders

The lambs from each ewe should be carefully noted so that when the sorting comes (usually June or July) it can be seen which ewes are breeding satisfactorily, and what class of ram suits them best, because possibly some of the most promising lambs may be the offspring of ewes that would otherwise be discarded. In a pure-bred flock, a regular system of sorting at a certain age cannot be followed with advantage, for in some cases it is wise to keep a ewe—a good ram breeder—as long as she will continue to breed, while others which produce nothing good as yearlings or two-year-olds may safely be put aside.

At the same time the breeder should try to continue a plan which keeps the flock from degenerating into a lot of old ewes.

As to the number of ewes which should be selected annually, one must be guided by circumstances. Should the young ewes be exceptionally good and by one or more sires which you have a high opinion of it will be wise to draw for the breeding flock more largely than usual. If on the other hand the yearling ewes are not to your liking it may be well to add none to the breeding flock but dispose of all the young ewes. These matters must be left to the judgment of the breeder but all such details are of great importance and whether they receive due attention or not means success or failure.

As before stated it is sound policy to manage the flock so it will not deteriorate into a lot of old worn out ewes, and with due care this can be avoided. A well-bred and good young flock must always have a far higher value than one in which several of its members have passed the prime.

The three C's in the clover-leaf represent Chandler Clover Hill Chariton. The distinct **quality** that Shropshires have which come from Chandler "Clover Hill" Chariton is widely known.

www.ingramcontent.com/pod-product-compliance
Lightning Source LLC
Chambersburg PA
CBHW081742220526

45468CB00008B/2204